LIFE AND WORKS OF AMBEDKAR

Dr. Shyam Shanker Prasad Gupta

CENTRUM PRESS
NEW DELHI-110002 (INDIA)

CENTRUM PRESS

H.O.: 4360/4, Ansari Road, Daryaganj,
New Delhi-110002 (India)
Tel: 23278000, 23261597, 23255577, 23286875

B.O.: No. 1015, Ist Main Road, BSK IIIrd Stage,
IIIrd Phase, IIIrd Block, Bangalore-560085 (INDIA)
Tel: 080-41723429

Email: centrumpress@gmail.com
Visit us at: www.centrumpress.com

Life and Works of Ambedkar

First Edition, 2011

ISBN 978-93-80836-47-8

PRINTED IN INDIA

Printed at Mehra Offset Press, Delhi

Contents

Preface

Dr. Ambedkar was the main architect of the Indian Constitution. He was born in a very poor low caste family of Madhya Pradesh. In U.S.A., he did his M.A. in 1915 and Ph.D. in 1916. From 1918 to 1920, he worked as a Professor of Law. Dr. Ambedkar set up his legal practice at the Mumbai High Court. Ambedkar was the main inspiration behind the inclusion of special provision in the Constitution of India for the development of Schedule Caste people. Dr. Ambedkar was the Law Minister of India from 1947 to 1951. He took part in the Satyagraha of untouchables at Nasik in 1930 for opening the Hindu temples to them.

Dr. Ambedkar was emancipator of the 'untouchables' and crusader for social justice. This liberator of the down trodden was affectionately called "Babasaheb". He was sthumously awarded 'Bharat Ratna' in the year 1990. Ambedkar was a great champion of the dalit cause because he succeeded in turning the depressed class movement into a revolutionary movement throughout India. Today India has witnessed the oppressed classes walking on the streets of cities and villages with confidence and poise, of course many despicable acts of discrimination and violence against the dalits still occur. Yet the juggernaut of equality is rolling on remorselessly and forcefully.

Ambedkar is India's foremost human rights activist during the 20th century. He is an emancipator, scholar, extraordinary social reformer and a true champion of human rights. It can be said that he is one of the highly regarded Indians whose emancipation and empowering role for oppressed groups that cut against the gender divide has inspired subaltern groups all over the world. All should try to take inspiration from Dr. B. R. Ambedkar's life and work for the creation of a just and gender-neutral world.

The book, which provides a rare opportunity to debate and discuss the contribution of Dr. Ambedkar in nation-building, will be great interest to all sections of people.

—Dr. Shyam Shankar Prasad Gupta

1

Introduction

Bhimrao Ramji Ambedkar (14 April 1891 — 6 December 1956), also known as *Babasaheb*, was an Indian jurist, political leader, Buddhist activist, philosopher, thinker, anthropologist, historian, orator, prolific writer, economist, scholar, editor, revolutionary and a revivalist for Buddhism in India. He was also the chief architect of the Indian Constitution. Born into a poor Mahar, then Untouchable, family, Ambedkar spent his whole life fighting against social discrimination, the system of *Chaturvarna* — the categorization of Hindu society into four *varnas* — and the Hindu caste system. He is also credited with providing a spark for the conversion of hundreds of thousands of Dalits with his Ambedkar (ite) Buddhism. Ambedkar has been honoured with the Bharat Ratna, India's highest civilian award.

Overcoming numerous social and financial obstacles, Ambedkar became one of the first "Dalit" to obtain a college education in India. Eventually earning law degrees and multiple doctorates for his study and research in law, economics and political science from Columbia University and the London School of Economics, Ambedkar returned home as a famous scholar and practiced law for a few years before publishing journals advocating political rights and social freedom for India's untouchables. He is regarded as a Bodhisattva by Indian Buddhists even though he never claimed himself to be a Bodhisattva.

Early Life

Bhimrao Ramji Ambedkar was born in the British-founded

town and military cantonment of Mhow in the Central Provinces (now in Madhya Pradesh). He was the 14th and last child of Ramji Maloji Sakpal and Bhimabai. His family was of Marathi background from the town of Ambavade in the Ratnagiri district of modern-day Maharashtra. They belonged to the Hindu, Mahar caste, who were treated as untouchables and subjected to intense socio-economic discrimination. Ambedkar's ancestors had for long been in the employment of the army of the British East India Company, and his father Ramji Sakpal served in the Indian Army at the Mhow cantonment. He had received a degree of formal education in Marathi and English, and encouraged his children to learn and work hard at school.

Belonging to the Kabir Panth, Ramji Sakpal encouraged his children to read the Hindu classics. He used his position in the army to lobby for his children to study at the government school, as they faced resistance owing to their caste. Although able to attend school, Ambedkar and other untouchable children were segregated and given no attention or assistance by the teachers. They were not allowed to sit inside the class. Even if they needed to drink water somebody from a higher caste would have to pour that water from a height as they were not allowed to touch either the water or the vessel that contained it. This task was usually performed for the young Ambedkar by the school peon, and if he could not be found Ambedkar went without water. Ramji Sakpal retired in 1894 and the family moved to Satara two years later. Shortly after their move, Ambedkar's mother died. The children were cared for by their paternal aunt, and lived in difficult circumstances. Only three sons — Balaram, Anandrao and Bhimrao — and two daughters — Manjula and Tulasa — of the Ambedkars would go on to survive them. Of his brothers and sisters, only Ambedkar succeeded in passing his examinations and graduating to a higher school. His native village name was "Ambavade" in Ratnagiri District so he changed his name from "Sakpal" to "Ambedkar" with the recommendation and faith of Mahadev Ambedkar, his Deshastha Brahmin teacher who believed in him.

Ramji Sakpal remarried in 1898, and the family moved to Mumbai (then Bombay), where Ambedkar became the first untouchable student at the Government High School near

Elphinstone Road. Although excelling in his studies, Ambedkar was increasingly disturbed by the segregation and discrimination that he faced. In 1907, he passed his matriculation examination and entered the University of Bombay, becoming one of the first persons of untouchable origin to enter a college in India. This success provoked celebrations in his community, and after a public ceremony he was presented with a biography of the Buddha by his teacher Krishnaji Arjun Keluskar also known as Dada Keluskar, a Maratha caste scholar. Ambedkar's marriage had been arranged the previous year as per Hindu custom, to Ramabai, a nine-year old girl from Dapoli. In 1908, he entered Elphinstone College and obtained a scholarship of twenty five rupees a month from the Gayakwad ruler of Baroda, Sahyaji Rao III for higher studies in the USA. By 1912, he obtained his degree in economics and political science, and prepared to take up employment with the Baroda state government. His wife gave birth to his first son, Yashwant, in the same year. Ambedkar had just moved his young family and started work, when he dashed back to Mumbai to see his ailing father, who died on February 2, 1913.

Fight against Untouchability

As a leading Indian scholar, Ambedkar had been invited to testify before the South borough Committee, which was preparing the Government of India Act 1919. At this hearing, Ambedkar argued for creating separate electorates and reservations for Dalits and other religious communities. In 1920, he began the publication of the weekly *Mooknayak* (*Leader of the Silent*) in Mumbai. Attaining popularity, Ambedkar used this journal to criticize orthodox Hindu politicians and a perceived reluctance of the Indian political community to fight caste discrimination. His speech at a Depressed Classes Conference in Kolhapur impressed the local state ruler Shahu IV, who shocked orthodox society by dining with Ambekdar. Ambedkar established a successful legal practice, and also organised the Bahishkrit Hitakarini Sabha to promote education and socio-economic uplifting of the depressed classes.

By 1927 Dr. Ambedkar decided to launch active movements against untouchability. He began with public movements and marches to open up and share public drinking water resources, also he began a struggle for the right to enter Hindu temples. He led a *satyagraha* in Mahad to fight for the right of the untouchable community to draw water from the main water tank of the town.

He was appointed to the Bombay Presidency Committee to work with the all-European Simon Commission in 1925. This commission had sparked great protests across India, and while its report was ignored by most Indians, Ambedkar himself wrote a separate set of recommendations for future constitutional reformers.

Poona Pact

By now Ambedkar had become one of the most prominent untouchable political figures of the time. He had grown increasingly critical of mainstream Indian political parties for their perceived lack of emphasis for the elimination of the caste system. Ambedkar criticized the Indian National Congress and its leader Mohandas Gandhi, whom he accused of reducing the untouchable community to a figure of pathos. Ambedkar was also dissatisfied with the failures of British rule, and advocated a political identity for untouchables separate from both the Congress and the British. At a Depressed Classes Conference on August 8, 1930 Ambedkar outlined his political vision, insisting that the safety of the Depressed Classes hinged on their being independent of the Government and the Congress both:

We must shape our course ourselves and by ourselves... Political power cannot be a panacea for the ills of the Depressed Classes. Their salvation lies in their social elevation. They must cleanse their evil habits. They must improve their bad ways of living.... They must be educated.... There is a great necessity to disturb their pathetic contentment and to instill into them that divine discontent which is the spring of all elevation. Ambedkar's political work had made him very unpopular with orthodox Hindus, as well as with many Congress politicians who had earlier condemned untouchability. This was largely because these "liberal" politicians usually stopped short of advocating full equality for untouchables.

In 1932, M. C. Rajah concluded a pact with two right-wingers in the Indian National Congress, Dr. B. S. Moonje and

Jadhav. According to this pact, Moonje offered reserved seats to scheduled castes in return for Rajah's support. This demand prompted Ambedkar to make an official demand for Separate Electorate System on an all-India basis. Ambedkar's prominence and popular support amongst the untouchable community had increased, and he was invited to attend the Second Round Table Conference in London in 1932. Gandhi fiercely opposed separate electorate for untouchables, Gandhi believed in Chaturvarna system, but accepted separate electorate for all other minority groups like Muslims, sikhs...etc. Gandhi feared that separate electorates for untouchables would divide Hindu society for future generations.

When the British agreed with Ambedkar and announced the awarding of separate electorates, Gandhi began a *fast-unto-death* while imprisoned in the Yerwada Central Jail of Pune in 1932 against the separate electorate for untouchables only. , Gandhi asked for the political unity of Hindus. Gandhi's fast provoked great public support across India, and orthodox Hindu leaders, Congress politicians and activists such as Madan Mohan Malaviya and Palwankar Baloo organized joint meetings with Ambedkar and his supporters at Yeravada. Fearing a communal reprisal and killings of untouchables in the event of Gandhi's death, Ambedkar agreed under massive coercion from the supporters of Gandhi . This agreement, which saw Gandhi end his fast, while dropping the demand for separate electorates that was promised through the British Communal Award prior to Ambedkar's meeting with Gandhi. Ambedkar was to later criticise this fast of Gandhi as a gimmick to deny political rights to the untouchables and increase the coercion he had faced to give up the demand for separate electorates.

Political Career

In 1935, Ambedkar was appointed principal of the Government Law College, Mumbai, a position he held for two years. Settling in Mumbai, Ambedkar oversaw the construction of a house, and stocked his personal library with more than 50,000 books. His wife Ramabai died after a long illness in the same year. It had been her long-standing wish to go on a pilgrimage to Pandharpur, but Ambedkar had refused to let her go, telling her that he would create a new Pandharpur for her instead of Hinduism's Pandharpur which treated them as untouchables. Speaking at the Yeola Conversion Conference on October 13 near Nasik, Ambedkar announced his intention to convert to a different religion and exhorted his followers to leave Hinduism. He would repeat his message at numerous public meetings across India

In 1936, Ambedkar founded the Independent Labour Party, which won 15 seats in the 1937 elections to the Central Legislative Assembly. He published his book *The Annihilation of Caste* in the same year, based on the thesis he had written in New York. Attaining immense popular success, Ambedkar's work strongly criticized Hindu orthodox religious leaders and the caste system in general. Ambedkar served on the Defence Advisory Committee and the Viceroy's Executive Council as minister for labour. With *What Congress and Gandhi Have Done to the Untouchables*, Ambedkar intensified his attacks on Gandhi and the Congress, hypocrisy. In his work *Who Were the Shudras?*, Ambedkar attempted to explain the formation of the Shudras i.e. the lowest caste in hierarchy of Hindu caste system. He also emphasised how Shudras are separate from Untouchables. Ambedkar oversaw the transformation of his political party into the All India Scheduled Castes Federation, although it performed poorly in the elections held in 1946 for the Constituent Assembly of India. In writing a sequel to *Who Were the Shudras?* in 1948, Ambedkar lambasted Hinduism in the *The Untouchables: A Thesis on the Origins of Untouchability*:

The Hindu Civilisation.... is a diabolical contrivance to suppress and enslave humanity. Its proper name would be infamy. What else can be said of a civilisation which has produced a mass of people... who are treated as an entity beyond human intercourse and whose mere touch is enough to cause pollution?

In a "communal malaise", both groups [Hindus and Muslims] ignore the urgent claims of social justice.

Pakistan or The Partition of India

Between 1941 and 1945, he published a number of books and pamphlets, including *Thoughts on Pakistan*, in which he criticized the Muslim League's demand for a separate Muslim state of Pakistan but considered its concession if Muslims demanded so as expedient.

In the above book Ambedkar wrote a sub-chapter titled *If Muslims truly and deeply desire Pakistan, their choice ought to be accepted.* He wrote that if the Muslims are bent on Pakistan, then it must be conceded to them. He asked whether Muslims in the army could be trusted to defend India. In the event of Muslims invading India or in the case of a Muslim rebellion, with whom would the Indian Muslims in the army side? He concluded that, in the interests of the safety of India, Pakistan should be acceded to, should the Muslims demand it. According to Ambedkar, the Hindu assumption that though Hindus and Muslims were two nations, they could live together under one state, was but a empty sermon, a mad project, to which no sane man would agree.

Father of India's Constitution

Upon India's independence on August 15, 1947, the new Congress-led government invited Ambedkar to serve as the nation's first law minister, which he accepted. On August 29, Ambedkar was appointed Chairman of the Constitution Drafting Committee, charged by the Assembly to write free India's new Constitution. Ambedkar won great praise from his colleagues and contemporary observers for his drafting work. In this task Ambedkar's study of sangha practice among early Buddhists and his extensive reading in Buddhist scriptures were to come to his aid. Sangha practice incorporated voting by ballot, rules of debate and precedence and the use of agendas, committees and proposals to conduct business. Sangha practice itself was modelled on the oligarchic system of governance followed by tribal republics of ancient India such as the Shakyas and the Lichchavis. Thus, although Ambedkar used Western models to give his Constitution shape, its spirit was Indian and, indeed, tribal.

Granville Austin has described the Indian Constitution drafted by Dr. Ambedkar as 'first and foremost a social document.' ... 'The majority of India's constitutional provisions are either directly arrived at furthering the aim of social revolution or attempt to foster this revolution by establishing conditions necessary for its achievement.'

The text prepared by Ambedkar provided constitutional guarantees and protections for a wide range of civil liberties for individual citizens, including freedom of religion, the abolition of untouchability and the outlawing of all forms of discrimination Ambedkar argued for extensive economic and social rights for women, and also won the Assembly's support for introducing a system of reservations of jobs in the civil services, schools and colleges for members of scheduled castes and scheduled tribes, a system akin to affirmative action. India's lawmakers hoped to eradicate the socio-economic inequalities and lack of opportunities for India's depressed classes through this measure, which had been originally envisioned as temporary on a need basis. The Constitution was adopted on November 26, 1949 by the Constituent Assembly.

Ambedkar resigned from the cabinet in 1951 following the stalling in parliament of his draft of the Hindu Code Bill, which sought to expound gender equality in the laws of inheritance, marriage and the economy. Although supported by Prime Minister Nehru, the cabinet and many other Congress leaders, it received criticism from a large number of members of parliament. Ambedkar independently contested an election in 1952 to the lower house of parliament, the Lok Sabha, but was defeated. He was appointed to the upper house, of parliament, the Rajya Sabha in March 1952 and would remain a member until his death.

Conversion to Buddhism

In the 1950s, Ambedkar turned his attention to Buddhism and travelled to Sri Lanka (then Ceylon) to attend a convention of Buddhist scholars and monks. While dedicating a new Buddhist vihara near Pune, Ambedkar announced that he was writing a book on Buddhism, and that as soon as it was finished, he planned to make a formal conversion to Buddhism. Ambedkar twice visited Burma in 1954; the second time in order to attend the third conference of the World Fellowship of Buddhists in Rangoon. In 1955, he founded the Bharatiya Bauddha Mahasabha, or the Buddhist Society of India. He completed his final work, *The Buddha and His Dhamma*, in 1956. It was published posthumously.

After meetings with the Sri Lankan Buddhist monk Hammalawa Saddhatissa, Ambedkar organised a formal public ceremony for himself and his supporters in Nagpur on October

14, 1956. Accepting the Three Refuges and Five Precepts from a Buddhist monk in the traditional manner, Ambedkar completed his own conversion. He then proceeded to convert an estimated 500,000 of his supporters who were gathered around him. Taking the 22 Vows. He then travelled to Kathmandu in Nepal to attend the Fourth World Buddhist Conference. His work on *The Buddha or Karl Marx* and "Revolution and counter-revolution in ancient India" (which was necessary for understanding his book "The Buddha and his dhamma") remained incomplete.

Death

Since 1948, Ambedkar had been suffering from diabetes. He was bed-ridden from June to October in 1954 owing to clinical depression and failing eyesight. He had been increasingly embittered by political issues, which took a toll on his health. His health worsened as he furiously worked through 1955. Just three days after completing his final manuscript *The Buddha and His Dhamma*, it is said that Ambedkar died in his sleep on December 6, 1956 at his home in Delhi.

Since the Caste Hindus denied the cremation at Dadar crematorium, A Buddhist-style cremation was organised for him at Chowpatty beach on December 7, attended by hundreds of thousands of supporters, activists and admirers. A conversion program was supposed to be organised on 16 December, 1956. So, those who had attended cremation function also got converted to buddhism at same place.

Ambedkar was survived by his second wife Savita Ambedkar and converted to Buddhism with him. His wife's name before marriage was Sharda Kabir. Savita Ambedkar died as a Buddhist in 2002. Ambedkar's grandson, Prakash Yaswant Ambedkar leads the Bharipa Bahujan Mahasangha and has served in both houses of the Indian Parliament.

A number of unfinished typescripts and handwritten drafts were found among Ambedkar's notes and papers and gradually made available. Among these were *Waiting for a Visa*, which probably dates from 1935–36 and is an autobiographical work, and the *Untouchables, or the Children of India's Ghetto*, which refers to the census of 1951. A memorial for Ambedkar was established in his Delhi house at 26 Alipur Road. His birthdate is celebrated as a public holiday known as Ambedkar Jayanti or Bhim Jayanti. He was posthumously awarded India's highest civilian honour, the Bharat Ratna in 1990. Many public institutions are named in his honour, such as the Dr. Babasaheb Ambedkar Open University in Hyderabad, Dr. BR Ambedkar University in Srikakulam Andhra Pradesh, B. R. Ambedkar Bihar University, Muzaffarpur Dr.B.R.Ambedkar National Institute of Technology, Jalandhar the other being Dr. Babasaheb Ambedkar International Airport in Nagpur, which was otherwise known as Sonegaon Airport. A large official portrait of Ambedkar is on display in the Indian Parliament building. On the anniversary of his birth (14 April) and death (6 December) and on Dhamma Chakra Pravartan Din, 14th Oct at Nagpur, at least half a million people gather to pay homage to him at his memorial in Mumbai. Thousands of bookshops are set up, and books are sold. His message to his followers was " Educate!!!, Agitate!!!, Organize!!!".

Dr. Babasaheb Ambedkar, Writings and Speeches

The Education Department, Government of Maharastra (Bombay) Published the Collection of Dr.B.R.Ambedkar's writings and speeches in different volumes.

Volume No.	Description
vol. 1.	Castes in India and 11 other essays
vol. 2.	Dr. Ambedkar in the Bombay Legislature, with the Simon Commission and at the Round Table Conferences, 1927-1939
vol. 3.	Philosophy of Hinduism ; India and the pre-requisites of communism ; Revolution and counter-revolution ;Buddha or Karl Marx
vol. 4.	Riddles in Hinduism
vol. 5.	Essays on untouchables and un-touchability
vol. 6.	The evolution of provincial finance in British India
vol. 7.	Who were the shudras? ; The untouchables
vol. 8.	Pakistan or the partition of India
vol. 9.	What Congress and Gandhi have done to the untouchables ; Mr. Gandhi and the emancipation of the untouchables

vol. 10. Dr. Ambedkar as member of the Governor General's Executive Council, 1942-46

vol. 11. The Buddha and his Dhamma

vol. 12. Unpublished writings ; Ancient Indian commerce ; Notes on laws ; Waiting for a Visa ; Miscellaneous notes, etc.

vol. 13. Dr. Ambedkar as the principal architect of the Constitution of India

vol. 14. (2 parts) Dr. Ambedkar as free India's first Law Minister and member of opposition in Indian Parliament (1947-1956)

vol. 16. Dr. Ambedkar's The Pali grammar

vol. 17. (3 parts) Dr. Babasaheb Ambedkar and his egalitarian revolution

vol. 18. (3 parts) Dr. Babasaheba Ambcdakara lekhana ani bhashane

Criticism and Legacy

Ambedkar's legacy as a socio-political reformer, had a deep effect on modern India. In post-Independence India his socio-political thought has acquired respect across the political spectrum. His initiatives have influenced various spheres of life and transformed the way India today looks at socio-economic policies, education and affirmative action through socio-economic and legal incentives. His reputation as a scholar led to his appointment as free India's first law minister, and chairman of the committee responsible to draft a constitution. He passionately believed in the freedom of the individual and criticised equally both orthodox casteist Hindu society. His condemnation of Hinduism and its foundation of caste system, made him controversial, although his conversion to Buddhism sparked a revival in interest in Buddhist philosophy in India and abroad.

Ambedkar condemned Gandhi's support for the caste system and perpetuating untouchability. Dr.Ambedkar warned people," Don't call Gandhi a saint. He is a seasoned politician. When everything else fails, Gandhi will resort to intrigue." "Don't fall under Gandhi's spell, he's not God... Mahatmas have come and Mahatmas have gone but untouchables have remained untouchables." Ambedkar's political philosophy has given rise to a large number of Dalit political parties, publications and workers' unions that remain active across India, especially in Maharashtra. His promotion of the Dalit Buddhist movement has rejuvenated interest in Buddhist philosophy in many parts of India. Mass conversion ceremonies have been organized by Dalit activists in modern times, emulating Ambedkar's Nagpur ceremony of 1956.

Some scholars, including some from the affected castes, took the view that the British were more even-handed between castes, and that continuance of British rule would have helped to eradicate many evil practices. This political opinion was shared by quite a number of social activists including Jyotirao Phule.

Some, in modern India, question the continued institution of reservations initiated by Ambedkar as outdated and anti-meritocratic. However, such arguments have always been dismissed by the Dalit masses. They express that the opposition of Caste-based reservations in India, primarily comes from the antagonism rooted in the Hindu society towards the Dalits. And, that the Caste-based reservations in India, in fact, have become the uplifting of Dalits in the post-colonial period.

Outside India, at the end of the 1990's, some Hungarian Romani people drew parallels between their own situation and the situation of the Dalits in India. Inspired by Ambedkar's approach, they started to convert to Buddhism.

In Popular Culture

Dr. Ambedkar's very name became a sign of victory of the down-trodden and long-exploited. *Jai Bhim*, i.e. *Victory to Bhim* has become a greeting phrase of the Buddhists all over in India.

Several movies, plays, and literary forms are made based on his life and teachings. Jabbar Patel directed the English-language movie (also dubbed in Hindi and other Indian languages) Dr. Babasaheb Ambedkar about the life of Ambedkar, released in 2000, starring the Indian actor Mammootty as Ambedkar. Sponsored by India's National Film Development Corporation and the Ministry of Social Justice, the film was released after a long and controversial gestation period. Mammootty won the National Film Award for Best

Actor for the role of Ambedkar, which he portrayed in this film. Dr. David Blundell, professor of anthropology at UCLA and Historical Ethnographer, has established a long-term project- a series of films and events that are intended to stimulate interest and knowledge about the social and welfare conditions in India. *Arising Light* is a film on the life on Dr. B. R. Ambedkar and social welfare in India.

Ambedkar Aur Gandhi, directed by Arvind Gaur and written by Rajesh Kumar, play tracks two prominent personalities of history — Mahatma Gandhi and Bhimraao Ambedkar.

Visions and Icons of Great Person

Every great person has a vision that impels all her/his works. Its discernibility may vary from case to case, generally being the function of the degree of turbulence around her/him, her/his relative position within the power structure in the given environment, her/his own equipment and conception of self-role. Marx, for instance, offers an articulate vision in clearest terms as he assumed the primary role of a philosopher to bring about revolutionary change, whereas Ambedkar had donned the mantle of mass-leadership in his primary role to spearhead the change; the degree of turbulence in the work domain of Marx had been minimal as he basically struggled in the realm of thought spanning complete human history whereas Ambedkar situated himself in the political turbulence that obtained in India as his strategy; Ambedkar's position in the power structure that bounded his work domain was certainly weak relative to Marx's.

This is neither to undermine the role of Marx as the activist constantly trying out his philosophy in the realm of practice nor to belittle the problems he suffered in life. With regard to personal equipment, both Ambedkar as well as Marx, could be taken to be equally equipped to undertake their respective tasks that they had undertaken. Marx had started off with philosophy and adopted the class-consciousness of the proletariat quite unlike Ambedkar, in whose case it was his own consciousness—the consciousness of an untouchable built up through concrete experience that had propelled his philosophical search.

Marx was well aware of his role in the revolutionary project, that he had to provide requisite tools and tackles for the working class for bringing about a change in the overall interest of humanity. But, Ambedkar was always loaded with anxiety as he had to strategize his way through the political maze around him, winning for dalits the maximum he could in a short span of time. In process, his role also underwent transformation with the expanse of the battleground. Inevitably, his thoughts and action always remained context-laden, polemical and pragmatically purposeful. It is therefore a relatively difficult task to discern a coherent vision underscoring the life work of Baba Saheb Ambedkar.

It is a moot point as to what extent a great person, who is essentially anchored in her/his space and time, could transcend these barriers and be equally effective in a different situation. A great person basically is the product of prevailing social relations. It is a particular moment in history that reflects an acute demand for such a person.

Depending upon her/his location in the social setting, she/he imparts her/his individual feature to the historical moments and movements in terms of working out specific means for resolving contradictions that engender them and releasing the forces of history in a specific direction. The masses whose cause she / he espouses throng around her/him in this process, depending upon the level of their collective consciousness. The longevity of the ideas a great person propounds in a historical setting depends upon the nature of contradictions, the size and expanse of problems and the time domain in which they are situated.

Generally, the classes that share the vision and ideology of such persons tend to iconise them with specific attributes of their class choice, in an attempt to institutionalise the latter. In this process, they would de-contextise some of the ideas and proffer them as universal theorems, if they perceive a pay-off for themselves in the sphere hegemonised by them.

This phenomenon becomes clear only over a long time horizon. For instance, the religious principles that were sprouted in the soil of certain specific social relations have basically blossomed in an alien soil with the help of the nutrients of class interests. Very broadly speaking, the trend of iconisation of great persons and the attempt of institutionalising their ideas is a gauge to assess the forces of status quo in the society.

Commitment and Constraint

In the case of Baba Saheb Ambedkar, iconisation was inevitable. The combination of factors like his high stature, his devotion to the cause of his people; the historical setting in which he lived, the low level of literacy and political consciousness in masses; and the vested interests of internal as well as external people have been its cause. The problem is not with iconisation as it is with its multiplicity. A question may be pertinently asked can Ambedkar be uniquely represented by a single icon? As Prof. Upendra Bakshi had outlined in one of his articles during the centenary year of his birth anniversary that there were many Ambedkars and had questioned as to which Ambedkar do we commemorate? When he said so, Prof. Bakshi was referring to different facets of Ambedkar's personality that could be virtually segregated.

One can even periodise some of them. For example, the pre-1942 Ambedkar as a young, untouchable man endowed with highest scholastic distinctions, struggling within and without for the emancipation of his people is a grossly different personality than the Ambedkar as a member of the viceroy's Executive Council or the Ambedkar as the law minister in the Nehru cabinet in the post-independence India or the Ambedkar as the chairman of the drafting committee for the Indian Constitution or even the Ambedkar of still later years who had completely identified himself with Buddhism and in a way completely spiritualised himself.

What comes clearly however, is that the changes in his outlook and role were essentially driven by his unstilted commitment to the cause of emancipation of oppressed humanity in general and dalits in particular. He might not have had appropriate methodological tools to deal with the problem at hand. With the equipment that basically belonged to a school of social engineers, he tried to dissect history. Paradoxically, he attempted to demolish the establishment with the very tools that were forged to serve the ruling classes. By training he did not have the facility to look at history as the continuum of human struggle with a certain inherent logic. He did use history as a repertoire of human episodes and attributed even logic to it but its source was externalised.

Non-Dialectical Solution: State and Religion

It appears that Baba Saheb Ambedkar had really internalised the doctrine of momentariness (*Anityatawad* and later *Kshanikwad*) of Buddha and therefore even refused to care for consistency in his views and opinions. This doctrine states that every thing changes every moment, that things are constantly becoming. It follows that in this situation of flux not even mental processes could be static, they had essentially to match the dynamicity of the material world. He thus never hesitated in changing his thoughts or strategy as per the unfolding situation. Viewed another way, these changes can be understood in relation to foci of control.

The degree of consistency in thought and action is generally inversely proportional to the distance of the subject from the foci of control of its surrounding. Ambedkar had nil or little control over his situation. He had to consistently create space for himself and strategize to influence the situation to his advantage. (The dynamics of the situation was propelled by the forces that were variously placed in the adversary camps.) The framework within which he conceived his struggle had exposed him to his lot to respond to this dynamics. The hallmark of Ambedkar's thoughts is the dynamic rationale, which he has consistently employed to comprehend situations and to strategize his response thereto. 'Ambedkar' therefore cannot be captured in static terms. His icon will have to represent the dynamism that he lived. Since, this is an infeasible proposition; we will have to discern the underscoring vision behind his works, the intransient essence of his entire mission to create a suitable icon. This icon, even if it does not resemble the familiar Ambedkar, alone could be the beacon of the Dalit movement.

The concept of *Anityawad* in Buddhism essentially belongs to dialectics that has made Buddha an early dialectician philosopher. The dichotomy that creeps in can only be resolved by dialectical method. It may be questioned whether Ambedkar's method was dialectical. It appears that while he accepts constant becoming of things as the principle underscoring the universe, he faces a dilemma with respect to the conception of order in this State. It could be resolved dialectically in terms of systemic attribute of self-regulation—a characteristic of internal control. But the conventional conception of order, essentially a non-dialectical conception, leads to externalisation of control. Ambedkar, having experienced the brutal aspects of history and unbridled exploitation of man by man, appears in need of a control mechanism operating at two levels, *viz.*, internal and external, so as to maintain the societal order in the desired State. His internal control mechanism is the moral code provided by the religion and the one for external control is the State.

If this moral code is internalised by all individuals and in turn by society as the summation of the latter (as the liberal tradition held), society is expected to have an internal order. If however the baser instincts of some people or group of people defy this order, either as a result of conflicting codes they follow or for any other reason, then in such case the State will step in and restore the order. The will of the collective is supposed to be embodied in the State by the Constitution. It is therefore that Ambedkar has reservation in agreeing with Marx that 'religion was the opium of masses' or the 'State shall eventually wither away'. Ambedkar certainly did not know that the order could be the attribute of the system itself. It is only in the sixties that Cybernetics principles came to lime light that the complex probabilistic systems, which the social systems certainly are, do have the inherent capability of self-regulating and self-organising control.

Chronological: the Struggle for an Education

Born in 1891, the young Ambedkar had (for his caste, for his time and place) a relatively comfortable upbringing. But his mother died when he was only five, and his early childhood brought other painful experiences as well, as he began to experience the full degradation of his place in the subbasement of the caste system.

1891, April 14: Bhimrao Ramji Ambedkar was born in the British-founded town of Mhow, an important military centre near Indore, Madhya Pradesh. He was the fourteenth and last child of Ramji Sankpal and Bhimabai Murbadkar Sankpal. The family's ancestral town was Ambavade (in the Ratnagiri District of Maharashtra).

1894: Gopal Baba Walangkar, retired and living in Dapoli, created the first public petition of the Untouchable movement: it asked the British colonial army to resume its recruitment of from the Untouchable castes.

1894-1896: When Bhimrao's father retired from his career with the British Army in 1894, he settled for a time in Dapoli (in Ratnagiri District). The young Bhimrao had his earliest education there: "At Dapoli in Bombay Presidency, however, there was a government-aided school, and the elder Ambedkar insisted his boys be allowed to attend on the ground that he was an army officer. It was finally arranged that they and four other "untouchables" might go to the school on the condition that they stay in a room by themselves and never come in contact with the caste children, and above all that they never take a drink from the school water supply. Those terms were accepted, and the future Doctor of Philosophy of Morningside Heights had his first conscious experience in ostracism and in learning at the same time. He was then 6 years old. The Hindu teacher at the school never entered the room in which the outcast children were struggling with their lessons. But occasionally he went to the door, whereupon the six small boys placed their slates on the ground, where he could see them, and then retreated to a far corner to listen to any comment the teacher might have to make. They could ask no questions. If they did not understand the lesson, there was no help from the teacher.... he was the only one of the group who got beyond the first school."

1896: The family moved to Satara, where Ramji Sakpal found a job with the Public Works Department in Goregaon; Bhimrao was enrolled in school in Satara.

1896: Bhimabai Sakpal died; of her fourteen children, only three sons (Balaram, Anandrao, Bhimrao) and two daughters (Manjula, Tulasa) survived her. The children were cared for by their paternal aunt Mira, who had a disabling hunchback but did her best to look after them. "Our family came originally from Dapoli Taluka of the Ratnagiri District of the Bombay

Presidency. From the very commencement of the rule of the East India Company, my forefathers had left their hereditary occupation for service in the Army of the Company. My father also followed the family tradition and sought service in the Army. He rose to the rank of an officer, and was a Subhedar when he retired. On his retirement my father took the family to Dapoli with a view to settling down there. But for some reason my father changed his mind. The family left Dapoli for Satara, where we lived till 1904."

"My father was a military officer, but at the same time a very religious person. He brought me up under a strict discipline. From my early age I found certain contradictions in my father's religious way of life. He was a Kabirpanthi, though his father was Ramanandi. As such, he did not believe in Murti Puja (Idol Worship), and yet he performed Ganapati Puja—of course for our sake, but I did not like it. He read the books of his Panth. At the same time, he compelled me and my elder brother to read every day before going to bed a portion of Mahabharata and Ramayana to my sisters and other persons who assembled at my father's house to hear the Katha. This went on for a long number of years."

A guarded gateway, but with a gap Bhimrao studied at Satara and then in Bombay, where he did so well on his exams that he was admitted to Bombay University—a unique feat for a member of the Mahar caste in his time and place. His marriage was arranged; amidst any amount of turmoil, he focused firmly on his studies.

Bhimrao entered the Government Middle School at Satara: "His second experience was at Satara, where he was the only untouchable pupil. He was allowed to sit in the same room with other boys, but always on the floor by himself in a remote corner. None could play with him or speak to him." "There was another Brahmin teacher in the High School. His surname was Ambedkar. Obliging and humane, he was a very irregular teacher. He loved Bhim very much. He dropped daily a part of his meal—boiled rice, bread and vegetables—into the hands of Bhim during recess. This teacher has left his impress on the life of his pupil. The original surname of Bhim's father was Sakpal.

It was a family name. Bhim drew his surname Ambedkar from his native village of Ambavade, as Maharashtrian surnames are often derived from the names of the ancestral villages. The teacher took so much fancy to the boy that he even changed his surname from Ambedkar to his own surname Ambedkar in the school records.... Ambedkar gratefully remembered this teacher."

"This incident gave me a shock such as I had never received before, and it made me think about untouchability—which, before this incident happened, was with me a matter of course, as it is with many touchables as well as the untouchables.".... "The incident, which I am recording as well as I can remember, occurred in about 1901, when we were at Satara. My mother was then dead. My father was away on service..."

1901: Ramji Sakpal, who had remarried in 1898, moved his family from Satara to Bombay, the capital city of Bombay Presidency. They found housing in the Dabak Chawl, Lower Parel. Bhimrao soon entered Elphinstone High School in Bombay: "At the age of 13 he went to the government high school at Elphinstone, becoming its [one] untouchable student. Here also he was ostracized, but was allowed to sit alone on a back bench. By this time his abilities with his lessons began to attract attention."

"One day it so happened that the class teacher called upon Bhim to come to the black board to solve an example. Instantaneously there was an uproar in the class. The caste Hindu children used to keep their tiffin-boxes behind the blackboard. Since they feared that their food would be polluted by Bhim's presence near the board, they dashed to the blackboard and hurled their tiffin-boxes aside before Bhim could reach and touch the blackboard. During his high school days both Bhim's elder brother and he were not allowed to take up Sanskrit as the second language. It was the key to the study of the Vedas which were neither to be heard nor to be read by the Sudras and the Atisudras—the Untouchables."

1902: The progressive-minded Shahu I (1884-1922), Maharaja of Kolhapur ordered 50% of the posts in the Kolhapur state services to be reserved for the backward classes. In 1907 he started two hostels open to Depressed Class boys.

1903: Shivram Janba Kamble, a Mahar from Poona, convened a meeting of Mahars from 51 villages at Saswad, near Poona; the result was a petition sent to the Governor of Bombay that requested admission into government jobs, public schools, the police, and the army.

1906: Bhimrao's marriage was arranged, with Ramabai, nine-year-old daughter of Bhiku Dhutre of Wanand, near Dapoli. Some accounts have it that she was related to Gopal Baba Walangkar.

1907: Bhimrao passed the Matriculation Examination that entitled him to enrol in a college affiliated with Bombay University; his marks were average, his best subject was Persian (which he studied in place of Sanskrit).

> *"My community people wanted to celebrate the occasion by holding a public meeting to congratulate me. Compared to the state of education in other communities, this was hardly an occasion for celebration. But it was felt by the organisers that I was the first boy in my community to reach this stage; they thought that I had reached a great height. They went to my father to ask for his permission. My father flatly refused, saying that such a thing would inflate the boy's head; after all, he has only passed an examination and done nothing more. Those who wanted to celebrate the event were greatly disappointed. They, however, did not give way. They went to Dada Keluskar, a personal friend of my father, and asked him to intervene. He agreed. After a little argumentation, my father yielded, and the meeting was held. Dada Keluskar presided. He was a literary person of his time. At the end of his address he gave me as a gift a copy of his book on the life of the Buddha, which he had written for the Baroda Sayajirao Oriental Series. I read the book with great interest, and was greatly impressed and moved by it."*

1908: Bhimrao entered Elphinstone College, a college affiliated with Bombay University.

1910: Shivram Janba Kamble, another early caste reformer, organized a second Mahar conference at Jejuri; this resulted in a memorandum sent to the British government. Young

Bhimrao met the reform-minded Gaikwar of Baroda, Sayaji Rao III (r.1875-1939) who then approved a scholarship of Rs. 25 a month for his education.

Off to Columbia, and on to London: The great escape! At the age of 22, the young Ambedkar came to Columbia, and began to make the intellectual and personal connections that shaped the rest of his life. He experienced what it was to be free—for a time—from the stigma of untouchability.

1912: Bhimrao passed the B.A. Examination (special subjects: Economics and Politics) from Bombay University, and prepared to take a position in the administration of Baroda State. His oldest son, Yashwant, was born.

1913: He had barely begun at his new post when he learned by telegram that his father was gravely ill; he rushed home just in time for a last farewell. "It was February 2, 1913, the saddest day in Bhimrao Ambedkar's life."

1913: The Gaikwar of Baroda announced his decision to offer scholarships to send students for higher education at Columbia University. A scholarship of 11.50 British pounds a month, for three years, was awarded to the young Ambedkar.

Receives Baroda State Scholarship to join the Political Science Department of the Columbia University as a Post Graduate Student where he worked under Professors Seligman, Clark, Seager, Moore, Mitchell, Chadwick, Simkovitch, Giddings, Dewey and Goldenweiser."

1913: Arriving in New York during the third week in July, Bhimrao was housed in Hartley Hall. But he didn't care for the food, and only stayed for a week. In August he moved from Hartley Hall to "Cosmopolitan Club" (554 West 114th Street), a housing club maintained by a group of Indian students. He finally settled in a dormitory, Livingston Hall (since renamed Wallach Hall, with his friend Naval Bhathena, a Parsi; the two remained friends for life.

"'The best friends I have had in life,' Dr. Ambedkar says, 'were some of my classmates at Columbia and my great professors, John Dewey, James Shotwell, Edwin Seligman, and James Harvey Robinson.'"

At Columbia: Prof. John Dewey: One of the major philosophers of education of the twentieth century, John Dewey

(1859-1952) became one of the young Ambedkar's heroes. Writing in 1936, Ambedkar referred to the work of "Prof. John Dewey, who was my teacher and to whom I owe so much."

Profs. Shotwell and Robinson: Another of the young Ambedkar's mentors, Prof. James Shotwell (1874-1965) was a Barnard historian who specialized in international relations, and a former student of Prof. James Harvey Robinson (1863-1936), Barnard's first historian—who himself was another of the mentors named by Dr. Ambedkar. Prof. Edwin Seligman: A friend of Lala Lajpat Rai, the well-known economist Edwin R. A. Seligman (1861-1939) became a particularly sympathetic mentor to the young Ambedkar, who continued to correspond with him for years.

During his three years at Columbia (including summers) 29 courses in economics, 11 in history, 6 in sociology, 5 in philosophy, 4 in anthropology, 3 in politics, and 1 each in elementary French and German. (Source: Office of the Registrar, Columbia University.)

"Parents can mold the destiny of children, and if we but follow this principle, be sure that we shall soon see better days; and our progress will be greatly accelerated if male education is pursued side by side with female education, the fruits of which you can very well see verified in your own daughter," Ambedkar wrote from New York in a Marathi letter to a friend of his father. "Let your mission therefore be to educate and preach the idea of education to those at least who are near to and in close contact with you."

1915: The young graduate student passed his M.A. exam in June, majoring in Economics, with Sociology, History, Philosophy, and Anthropology as other subjects of study; he presented a thesis, "Ancient Indian Commerce". For his outstanding achievement, he was honoured by students and professors of the Faculty of Arts at a special dinner. In 1916 he offered another M.A. thesis, "National Dividend of India—A Historic and Analytical Study"; it was this one that later became the nucleus of his Ph.D. dissertation.

1916: On May 9th, he read his paper "Castes in India: Their Mechanism, Genesis, and Development" before a seminar conducted by the anthropologist Prof. Alexander Goldenweiser

(1880-1940). Dr. Ambedkar was very proud of this paper, and remained so. He promptly got it published in the Indian Antiquary (May 1917). As late as 1936 he wrote that only shortage of time prevented him from reworking Annihilation of Caste so as to include in it this early seminar paper.

1916: In June he went to London, and in October he was admitted to Gray's Inn for Law, and to the London School of Economics and Political Science for Economics, where he was allowed to start work on a doctoral thesis. He often worked in the British Library Reading Room.

1917: The term of his scholarship from Baroda ended, so that he was obliged to come back to India in June with his work unfinished; he was, however, given permission to return and finish within four years. He sent his precious and much-loved collection of books back on a steamer—but it was torpedoed and sunk by a German submarine.

He was appointed Military Secretary to the Gaikwar of Baroda; he had agreed to join the Baroda service as a condition of his scholarship. But this experience was not a happy one. Even to reach Baroda, he had to pay his own expenses; he used the damages paid by Thomas Cook and Company for his torpedoed luggage. And when he arrived in Baroda, things went from bad to worse:

> *"My five years of staying in Europe and America had completely wiped out of my mind any consciousness that I was an untouchable, and that an untouchable wherever he went in India was a problem to himself and to others. But when I came out of the station, my mind was considerably disturbed by a question, 'Where to go? Who will take me?'*

1917: Meeting in Calcutta with Annie Besant as its President, for the first time in its history the Indian National Congress adopted a resolution endorsing "the justice and righteousness of removing all disabilities imposed by custom upon the Depressed Classes."

1918: After the Baroda fiasco, he tried to find ways to make a living for his growing family. With the help of Parsi friends, he became a private tutor, and found some work as an accountant. He also started an investment consulting business, but it failed when his clients learned that he was an untouchable. Finally he became Professor of Political Economy in the Sydenham College of Commerce and Economics, in Bombay. (This position came about through the recommendation of his London acquaintance, Lord Sydenham, former Governor of Bombay.) He was mostly successful with his students, but some of the other professors objected to his sharing the same drinking-water jug that they all used.

In the new *Journal of Indian Economics* (1-1), he reviewed a book by Bertrand Russell: "Mr. Russell and the Reconstruction of Society". And in the new *Journal of the Indian Economic Society* (1,2-3) he published "Small Holdings in India and Their Remedies".

1919: He testified both orally and in writing before the Southborough Committee, which was investigating franchise matters in the light of the planned Montagu-Chelmsford reforms. He demanded separate electorates and reserved seats for the untouchables: "The real social divisions of India then are: (1) Touchable Hindus. (2) Untouchable Hindus. (3) Mohammedans. (4) Christians. (5) Parsees.

1920: Dr. Ambedkar started a weekly paper, "Mooknayak" ("Leader of the Voiceless"), in Marathi, with the help of the reform-minded Shahu I (1884-1922), Maharaja of Kolhapur. In the first issue he called India a "home of inequality," and described Hindu society as "a tower which had several storeys without a ladder or an entrance. One was to die in the storey in which one was born." The Depressed Classes must be saved "from perpetual slavery, poverty, and ignorance"; herculean efforts must be made "to awaken them to their disabilities."

In March, he spoke at a Depressed Classes conference in Mangaon in Kolhapur State; it was attended by the Maharaja of Kolhapur, who publicly praised him as a future national leader. At the end of the conference the Maharaja and his courtiers shocked the tradition-minded by actually dining with Ambedkar and his caste members.

In May, the Maharajah of Kolhapur convened another such conference, in Nagpur, a town later to acquire a major symbolic significance in Dr. Ambedkar's life. "At the conclusion of the conference, Ambedkar made an attempt in the direction of consolidating the forces of the Depressed Classes. In the Central Provinces the Mahar community had eighteen sub-castes. He called the leaders of the community together and gave a dinner in which they all participated.

It should be noted that with great persuasion Ambedkar could get all the sub-castes of the Mahar community, and not all the Untouchable communities, to dine together. It was not possible yet to make all the communities belonging to the Untouchables participate in an intercaste dinner!"

Having resigned from his teaching position, in July he returned to London, relying on his own savings, supplemented by loans from the Maharaja of Kolhapur and his friend Naval Bhathena. He returned to the London School of Economics, and to Gray's Inn to read for the Bar. He lived in poverty, and studied constantly in the British Museum.

1922: Through unremitting hard work, Ambedkar once again over fulfilled all expectations: he completed a thesis for a M.Sc. (Econonics) degree at London School of Economics, and was called to the bar, and submitted a Ph.D. thesis in economics to the University of London.

1923: His Ph.D. thesis at the University of London, "The Problem of the Rupee", was challenged on political grounds (for its allegedly subversive, anti-British implications), but was resubmitted and finally accepted; it was at once published in London (by P. S. King and Son, Ltd.), and was "dedicated to the memory of my father and mother, as a token of my abiding gratitude for the sacrifices they made and the enlightenment they showed in the matter of my education."

1924: Back in India, Dr. Ambedkar began to practice as a barrister in Bombay, and also began to lecture part-time at Batliboi's Accountancy Training Institute. He founded the "Bahishkrit Hitakarini Sabha" (Group for the Wellbeing of the Excluded), to help the Depressed Classes mobilize. Its motto was "Educate, Agitate, Organise."

1925: He published his London School of Economics M.A. thesis as "The Evolution of Provincial Finance in British India", it was dedicated to the Gaikward of Baroda ("for his help in the matter of my education"), and had an introduction by Prof.

Seligman. He also gave testimony before the Royal Commission on Indian Currency and Finance.

"There is no use pretending that I and my wife have recovered from the shock of our son's death, and I do not think that we ever shall. We have in all buried four precious children, three sons and a daughter, all lively, auspicious, and handsome children. The thought of this is sufficiently crushing, let alone the future which would have been theirs if they had lived....My last boy was a wonderful boy, the like of whom I have seldom seen. With his passing away life to me is a garden full of weeds. But enough of this, I am too overcome to write any more."—A letter to a friend, Aug. 16, 1926.

1926: The Governor of Bombay nominated him as a member of the Bombay Legislative Council; he took his duties seriously, and often delivered speeches on economic matters.

He led the satyagraha at Mahad to exercise the right of Untouchables to draw water from the Chavdar Tank. He ceremonially took a drink of water from the tank, after which local caste Hindus rioted, and Brahmins took elaborate measure for the ritual purification of the tank.

1927: On January 1st, he held a meeting at the Koregaon Victory Memorial, 17 miles from Poona, which commemorates the defeat of the Peshwa's forces and the inauguration of British rule. The names of Mahar soldiers who fought with the British are inscribed there on a marble tablet. Such meetings still take place annually there on that day.

On June 8, he was formally awarded the Ph.D. degree from Columbia University. His Ph.D. thesis was "The Evolution of Provincial Finance in British India".

On December 24th, he addressed a second Depressed Classes Conference in Mahad; he attacked the *Laws of Manu*, and then a copy of this ancient text was publicly burned, to the shock and horror of many caste Hindus.

1928: Dr. Ambedkar was appointed Professor at the Government Law College, Bombay; his term of appointment ended in 1929.

Dr. Ambedkar was selected as a member of the Bombay Presidency Committee to work with the Simon Commission, drafting guidelines for political change in India. Congress decided to boycott the Simon Commission because it had no Indians on it. Dissenting from the views of many of his colleagues, Dr. Ambedkar prepared a detailed report setting out his own recommendations.

1929: Dr. Ambedkar closed his second journal, "Bahiskrit Bharat" ("Excluded India"), which he had started in 1927, and replaced it with the "Janata" ("The People"), which was published until 1956, when it took on the name "Prabuddha Bharata" (after his conversion).

On Oct. 23, during a visit to Chalisgaon, he had a bad accident, and was confined to bed until the last week of December:

> *"At Chalisgaon I got down to go to a village on the Dhulia line, to investigate a case of social boycott which had been declared by the caste Hindus against the untouchables of that village. The untouchables of Chalisgaon came to the station and requested me to stay for the night with them..."*

The year was 1929. The Bombay Government had appointed a Committee to investigate the grievances of the untouchables. I was appointed a member of the Committee. The Committee had to tour all over the province to investigate the allegations of injustice, oppression and tyranny. The Committee split up. I and another member were assigned the two districts of Khandesh. My colleague and myself, after finishing our work, parted company. He went to see some Hindu saint. I left by train to go to Bombay. At Chalisgaon I got down to go to a village on the Dhulia line, to investigate a case of social boycott which had been declared by the caste Hindus against the untouchables of that village. The untouchables of Chalisgaon came to the station and requested me to stay for the night with them. My original plan was to go straight to Bombay after investigating the case of social boycott. But as they were keen, I agreed to stay overnight. I boarded the train for Dhulia to go to the village, went there and informed myself of the situation prevailing in the village, and returned by the next train to Chalisgaon. I found the untouchables of Chalisgaon waiting for me at the station. I was garlanded. The Maharwada, the quarters of the untouchables, is about two miles from the Railway Station, and to reach it one has to cross a river on which there is a culvert. There were many horse carriages at the station plying [available] for hire. The Maharwada was also within walking distance from the station. I expected immediately to be taken to the Maharwada. But there was no movement in that direction, and I could not understand why I was kept waiting.

After an hour or so a *tonga* (one-horse carriage) was brought close to the platform, and I got in. The driver and I were the only two occupants of the tonga. Others went on foot by a short cut. The tonga had not gone 200 paces when there was almost a collision with a motor car. I was surprised that the driver, who was hired for driving every day, should have been so inexperienced. The accident was averted only because on the loud shout of the policeman the driver of the car pulled it back.

We somehow came to the culvert on the river. On it there are no walls as there are on a bridge. There is only a row of stones fixed at a distance of five or ten feet. It is paved with stones. The culvert on the river is at right angles to the road we were coming by. A sharp turn has to be taken to come to the culvert from the road. Near the very first side stone of the culvert, the horse, instead of going straight, took a turn and bolted. The wheel of the tonga struck against the side stone so forcibly that I was bodily lifted up and thrown down on the stone pavement of the culvert, and the horse and the carriage fell down from the culvert into the river.

So heavy was the fall that I lay down [=there] senseless. The Maharwada is just on the other bank of the river. The men who had come to greet me at the station had reached there ahead of me. I was lifted and taken to the Maharwada amidst the cries and lamentations of the men, women and children. As a result of this I received several injuries. My leg was fractured, and I was disabled for several days. I could not understand how all this had happened. The tongas pass and repass the culvert every day, and never has a driver failed to take the tonga safely over the culvert.

On enquiry I was told the real facts. The delay at the railway station was due to the fact that the tongawalas were not prepared to drive the tonga with a passenger who was an untouchable. It was beneath their dignity. The Mahars could not tolerate that I should walk to their quarters. It was not in keeping with their sense of my dignity. A compromise was therefore arrived at. That compromise was to this effect: the owner of the tonga would give the tonga on hire, but not drive. The Mahars may [could] take the tonga, but must find someone to drive it.

The Mahars thought this to be a happy solution. But they evidently forgot that the safety of the passenger was more important than the maintenance of his dignity. If they had thought of this, they would have considered whether they could get a driver who could safely conduct me to my destination. As a matter of fact none of them could drive, because it was not their trade. They therefore asked someone from amongst themselves to drive. The man took the reins in his hand and started, thinking there was nothing in it. But as he got on, he felt his responsibility and became so nervous that he gave up all attempt to control. To save my dignity, the Mahars of Chalisgaon had put my very life in jeopardy. It was then I learnt that a Hindu tongawalla, no better than a menial, has a dignity by which he can look upon himself as a person who is superior to any untouchable, even though he may be a Barrister-at-law.

In the year 1934, some of my co-workers in the movement of the depressed classes expressed a desire to go on a sightseeing tour, if I agreed to join them. I agreed. It was decided that our plan should at all events include a visit to the Buddhist caves at Verul. It was arranged that I should go to Nasik, and the party should join me at Nasik. To go to Verul we had to go to Aurangabad. Aurangabad is a town in the Mohammedan State of Hyderabad, and is included in the dominion of His Exalted Highness, the Nizam.

On the way to Aurangabad we had first to pass another town called Daulatabad, which is also in the Hyderabad State. Daulatabad is a historical place and was, at one time, the capital of a famous Hindu King, by name Ramdeo Rai. The fort of Daulatabad is an ancient historical monument,, and no tourist while in that vicinity should omit a visit to it. Accordingly our party had also included in its programme a visit to the fort of Daulatabad.

We hired some buses and touring cars. We were about thirty in number. We started from Nasik to Yeola, as Yeola is on the way to Aurangabad. Our tour programme had not been announced—and quite deliberately. We wanted to travel incognito, in order to avoid difficulties which an untouchable tourist has to face in outlying parts of the country. We had informed only our people at those centres at which we had decided to halt. Accordingly, although on the way we passed many villages in the Nizam's State, none of our people had come to meet us. It was naturally different at Daulatabad. There our people had been informed that we were coming. They were waiting for us and had gathered at the entrance to the town. They asked us to get down and have tea and refreshment first, and then to go to see the fort. We did not agree to their proposal. We wanted tea very badly, but we wanted sufficient time to see the fort before it was dusk. We therefore left for the fort, and told our people that we would take tea on our return. Accordingly we told our drivers to move on, and within a few minutes we were at the gate of the fort.

The month was Ramadan, the month of fast for the Mohammedans. Just outside the gate of the fort there is a small tank of water full to the brim. There is all around a wide stone pavement. Our faces, bodies and clothes were full of dust gathered in the course of our journey, and we all wished to have a wash. Without much thought, some members of the party washed their faces and their legs on the pavement with the water from the tank. After these ablutions, we went to the gate of the fort. There were armed soldiers inside. They opened the big gates and admitted us into the archway.

We had just commenced asking the guard the procedure for obtaining permission to go into the fort. In the meantime an old Mohammedan with [a] white flowing beard was coming from behind shouting "The Dheds (meaning untouchables) have polluted the tank!" Soon all the young and old Mohammedans who were near about joined him and all started abusing us. "The Dheds have become arrogant. The Dheds have forgotten their religion (*i.e.* to remain low and degraded). The Dheds must be taught a lesson." They assumed a most menacing mood. We told them that we were outsiders and did not know the local custom. They turned the fire of their wrath against the local untouchables, who by that time had arrived at the gate. "Why did you not tell these outsiders that this tank could not be used by untouchables?" was the question they kept on asking them. Poor people! They were not there when we entered [the] tank [area]. It was really our mistake, because we acted without inquiry. The local untouchables protested that it was not their fault.

But the Mohammedans were not prepared to listen to my explanation. They kept on abusing them and us. The abuse was so vulgar that it exasperated us. There could easily have been a riot, and possibly murders. We had, however, to restrain ourselves. We did not want to be involved in a criminal case which would bring our tour to an abrupt end. One young Muslim in the crowd kept on saying that everyone must conform to his religion, meaning thereby that the untouchables must not take water from a public tank. I had grown quite impatient, and asked him in a somewhat angry tone, "Is that what your religion teaches? Would you prevent an untouchable from taking water from this tank if he became a Mohammedan?" These straight questions seemed to have some effect on the Mohammedans. They gave no answer, and stood silent.

Turning to the guard I said, again in an angry tone, "Can we get into the fort or not? Tell us; if we can't, we don't want to stay." The guard asked for my name. I wrote it out on a piece of paper. He took it to the Superintendent inside, and came out. We were told that we could go into the fort, but we could not touch water anywhere in the fort; and an armed soldier was ordered to go with us to see that we did not transgress the order.

I gave one instance to show that a person who is an untouchable to a Hindu is also an untouchable to a Parsi. This will show that a person who is an untouchable to a Hindu is also an untouchable to a Mohammedan.

The next case is equally illuminating. It is a case of an Untouchable school teacher in a village in Kathiawar, and is reported in the following letter which appeared in the Young India, a journal published by Mr. Gandhi, in its issue of 12th December 1929. It expresses the difficulties he [=the writer] had experienced in persuading a Hindu doctor to attend to his wife, who had just delivered, and how the wife and child died for want of medical attention. The letter says:

> *"On the 5th of this month a child was born to me. On the 7th, she [the writer's wife] fell ill and suffered from loose stools. Her vitality seemed to ebb away and her chest became inflamed. Her breathing became difficult and there was acute pain in the ribs. I went to call a doctor—but he said he would not go to the house of a Harijan, nor was he prepared to examine the child. Then I went to [the] Nagarseth and Garasia Darbar and pleaded [with] them to help me. The Nagarseth stood surety to the doctor for my paying his fee of two rupees. Then the doctor came, but on condition that he would examine them only outside the Harijan colony. I took my wife out of the colony along with her newly born child. Then the doctor gave his thermometer to a Muslim, he gave it to me, and I gave it to my wife and then returned it by the same process after it had been applied. It was about eight o'clock in the evening and the doctor, on looking at the thermometer in the light of a lamp, said that the patient was suffering from pneumonia. Then the doctor went away and sent the medicine. I brought some linseed from the bazar and used it on the patient. The doctor refused to see her later, although I gave the two rupees fee. The disease is dangerous and God alone will help us.*

The lamp of my life has died out. She passed away at about two o'clock this afternoon." The name of the Untouchable school teacher is not given. So also the name of the doctor is not mentioned. This was at the request of the Untouchable teacher, who feared reprisals. The facts are indisputable. No explanation is necessary. The doctor, in spite of being educated, refused to apply the thermometer and treat an ailing woman in a critical condition. As a result of his refusal to treat her, the woman died. He felt no qualms of conscience in setting aside the code of conduct which is binding on his profession. The Hindu would prefer to be inhuman rather than touch an Untouchable.

There is one other incident more telling than this. On the 6th of March 1938, a meeting of the Bhangis was held at Kasarwadi (behind Woollen Mills), Dadar, Bombay, under the

Chairmanship of Mr. Indulal Yadnik. In this meeting, one Bhangi boy narrated his experience in the following terms:

> *"I passed the Vernacular Final Examination in 1933. I have studied English up to the 4th Standard. I applied to the Schools Committee of the Bombay Municipality for employment as a teacher, but I failed, as there was no vacancy. Then I applied to the Backward Classes Officer, Ahmedabad, for the job of a Talati (village Patwari), and I succeeded. On 19th February 1936, I was appointed a Talati in the office of the Mamlatdar of the Borsad Taluka in the Kheda District."*

Although my family originally came from Gujarat, I had never been in Gujarat before. This was my first occasion to go there. Similarly, I did not know that untouchability would be observed in Government Offices. Besides in my application the fact of my being a Harijan was mentioned and so I expected that my colleagues in the office would know beforehand who I was. That being so, I was surprised to find the attitude of the clerk of the Mamlatdar's office when I presented myself to take charge of the post of the Talati.

The Karkun contemptuously asked, "Who are you?" I replied, "Sir, I am a Harijan." He said, "Go away, stand at a distance. How dare you stand so near me! You are in office, if you were outside I would have given you six kicks. What audacity to come here for service!" Thereafter, he asked me to drop on the ground my certificate and the order of appointment as a Talati. He then picked them up. While I was working in the Mamlatdar's office at Borsad I experienced great difficulty in the matter of getting water for drinking. In the verandah of the office there were kept cans containing drinking water. There was a waterman in charge of these water cans. His duty was to pour out water to clerks in office whenever they needed it. In the absence of the waterman they could themselves take water out of the cans and drink it.

That was impossible in my case. I could not touch the cans, for my touch would pollute the water, I had therefore to depend upon the mercy of the waterman. For my use there was kept a small rusty pot No one would touch it or wash it except myself. It was in this pot that the waterman would dole out water to me. But I could get water only if the waterman was present. This waterman did not like the idea of supplying me with water. Seeing that I was coming for water, he would manage to slip away, with the result that I had to go without water; and the days on which I had no water to drink were by no means few.

I had the same difficulties regarding my residence. I was a stranger in Borsad. No caste Hindu would rent a house to me. The Untouchables of Borsad were not ready to give me lodgings, for the fear of displeasing the Hindus who did not like my attempt to live as a clerk, a station above me. Far greater difficulties were with regard to food. There was no place or person from where I could get my meals. I used to buy 'Bhajhas' morning and evening, eat them in some solitary place outside the village, and come and sleep at night on the pavement of the verandahs of the Mamlatdar's office. In this way, I passed four days. All this became unbearable to me. Then I went to live at Jentral, my ancestral village. It was six miles from Borsad. Every day I had to walk eleven miles. This I did for a month and a half.

Thereafter the Mamlatdar sent me to a Talati to learn the work. This Talati was in charge of three villages, Jentral, Khapur and Saijpur. Jentral was his headquarters. I was in Jentral with this Talati for two months. He taught me nothing, and I never once entered the village office. The headman of the village was particularly hostile. Once he had said, "Your fellows, your father, your brother are sweepers who sweep the village office, and you want to sit in the office as our equal? Take care, better give up this job!" One day the Talati called me to Saijpur to prepare the population table of the village. From Jentral I went to Saijpur. I found the Headman and the Talati in the village office doing some work. I went, stood near the door of the office, and wished them "good morning," but they took no notice of me. I stood outside for about fifteen minutes. I was already tired of life, and felt enraged at being thus ignored and insulted. I sat down on a chair that was lying there. Seeing me seated on the chair, the Headman and the Talati quietly went away without saying anything to me.

A short while after, people began to come, and soon a large crowd gathered round me. This crowd was led by the Librarian of the village library. I could not understand why an educated person should have led this mob. I subsequently learnt that the chair was his. He started abusing me in the worst terms. Addressing the Ravania (village servant) he said, "Who allowed this dirty dog of a Bhangi to sit on the chair?" The Ravania unseated me and took away the chair from me. I sat on the ground.

Thereupon the crowd entered the village office and surrounded me. It was a furious crowd raging with anger, some abusing me, some threatening to cut me to pieces with the Dharya (a sharp weapon like the sword). I implored them to excuse me and to have mercy upon me. That did not have any effect upon the crowd.

I did not know how to save myself. But an idea came to me of writing to the Mamlatdar about the fate that had befallen me, and telling him how to dispose of my body in case I was killed by the crowd. Incidentally, it was my hope that if the crowd came to know that I was practically reporting against them to the Mamlatdar, they might hold their hands. I asked the Ravania to give me a piece of paper, which he did. Then with my fountain pen I wrote the following on it in big bold letters so that everybody could read it:

"To,

The Mamlatdar, Taluka Borsad.

Sir,

Be pleased to accept the humble salutations of Parmar Kalidas Shivram. This is to humbly inform you that the hand of death is falling upon me today. It would not have been so if I had listened to the words of my parents. Be so good as to inform my parents of my death."

The Librarian read what I wrote and at once asked me to tear it off, which I did. They showered upon me innumerable insults. "You want us to address you as our Talati? You are a Bhangi and you want to enter the office and sit on the chair?" I begged for mercy and promised not to repeat this, and also promised to give up the job. I was kept there till seven in the evening, when the crowd left. By then the Talati and the Mukhiya had still not come. Thereafter I took fifteen days' leave and returned to my parents in Bombay."

Conflict, Controversy and Congress

Dr. Ambedkar was now in the midst of his career; this was the central and perhaps most controversy-filled decade of his whole complex life. He was often at odds with Congress, and was attacked by the nationalist press as a traitor. But as always, through all difficulties and frustrations, he persevered.

1930: On Aug. 8, Dr. Ambedkar presided over the Depressed Classes Congress at Nagpur, and made a major speech: he endorsed Dominion status, and criticized Gandhi's Salt March and civil disobedience movement as inopportune; but he also criticized British colonial misgovernment, with its famines and immiseration. He argued that the "safety of the Depressed Classes" hinged on their "being independent of the Government and the Congress" both: "We must shape our course ourselves and by ourselves." His conclusion emphasized self-help: "Political power cannot be a panacea for the ills of the Depressed Classes. Their salvation lies in their social elevation. They must cleanse their evil habits. They must improve their bad ways of living.... They must be educated.... There is a great necessity to disturb their pathetic contentment and to instil into them that divine discontent which is the spring of all elevation."

Dr. Ambedkar was invited by the Viceroy to be a delegate to the Round Table Conference, and left for London in October. He participated extensively in the work of the Round Table Conference, often submitting written statements of his views. His views at the time were described in an unpublished manuscript later found among his papers: "The Untouchables and the Pax Britannica".

"Prince and Outcast at Dinner in London end Age-old Barrier: Gaikwad of Baroda is Host to 'Untouchable' and Knight of High Hindu Caste.."

"But I tell you that the Congress is not sincere about its professions. Had it been sincere, it would have surely made the removal of untouchability a condition, like the wearing of khaddar, for becoming a member of the Congress." On August 14th, 1931, Dr. Ambedkar met with Gandhi for the first time. From Gandhi's side, their discussion was an absent-minded rebuke that seemed to be more in sorrow than in anger; from Ambedkar's side, it was an outburst of passionate reproach.

1932: The All-Indian Depressed Classes Conference, held at Kamtee near Nagpur on May 6th, backed Dr. Ambedkar's demand for separate electorates, rejecting compromises proposed by others. Gandhi, in Yeravda jail, started a fast to the death against the separate electorates granted to the Depressed Classes by Ramsay McDonald's Communal Award. By September 23, a very reluctant Dr. Ambedkar was obliged by the pressure of this moral blackmail to accept representation through joint electorates instead. The result was the Poona Pact. In 1933, Gandhi replaced his journal "Young India" with a new one called "Harijan," and undertook a 21-day "self-purification fast" against untouchability.

1933: Dr. Ambedkar participated in the work of the "Joint Committee on Indian Legislative Reform", examining a number of significant witnesses.

1935: Dr. Ambedkar was appointed Principal of the Government Law College, and became a professor there as well; he held these positions for two years. In May, Dr. Ambedkar's wife Ramabai died after a long illness. Her great wish had been to make a pilgrimage to Pandharpur, but since as an untouchable she would not have been allowed to enter the temple, her husband had never allowed her to go.

On Oct. 13th, Dr. Ambedkar presided over the Yeola Conversion Conference, held in Yeola, in Nasikh District. He advised the Depressed Classes to abandon all agitation for temple-entry privileges; instead, they should leave Hinduism entirely and embrace another religion. He vowed, "I solemnly assure you that I will not die as a Hindu." The struggle for social justice began to receive increasing attention and support from progressive writers. Mulk Raj Anand's powerful novel "Untouchable" (1935) was followed by "Coolie" (1936), with a foreword by E. M. Forster; both works called international attention to caste and class injustices. In Hindi, there was the work of Premchand. In December, Dr. Ambedkar was invited by the Jat-Pat-Todak Mandal of Lahore, a caste-reform organization, to preside over its annual conference in the spring of 1936.

1935/36: He composed (or began to compose?), but did not publish, a brief, moving, and largely autobiographical memoir called Waiting for a Visa.

1936: On April 13-14th, he addressed the Sikh Mission Conference in Amritsar, and reiterated his intention of renouncing Hinduism.In late April, the Jat-Pat-Todak Mandal realized the radical nature of its guest's planned speech, and withdrew its earlier invitation. On May 15th, Dr. Ambedkar published the speech he would have given, with an introductory account of the whole controversy. The result, a slim little book called "The Annihilation of Caste", became famous at once.

On May 31st, Dr. Ambedkar addressed the Mumbai Elaka Mahar Parishad (Bombay Mahar Society), during a meeting at Naigaum (Dadar), in Bombay. He spoke in Marathi, to his own people, with vividness and poignancy: "What Path to Salvation?". This was the only time he addressed an audience expressly limited to Mahars.

In August, he founded his first political party, the Independent Labour Party, which contested 17 seats in the 1937 General Elections, and won 15.

The Maharaja of Travancore issued a proclamation allowing temple entry to the Depressed Classes; this was the first such event in modern India.

1937: Dr. Ambedkar published the second edition of "The Annihilation of Caste", adding a concluding appendix that featured a debate with Gandhi over the speech. This work remained a best-seller, going through many editions in the coming years—and exciting much controversy. "It was logic on fire, pinching and pungent, piercing and fiery, provocative and explosive."

1938: Over Dr. Ambedkar's vigorous protests, in January Congress adopted Gandhi's own term "Harijans" ("Children of God") as the official name for the "scheduled castes." In protest against a term that he considered condescending and meaningless, Dr. Ambedkar and his party staged a walkout from the Bombay Legislative Assembly. Dr. Ambedkar made a number of significant speeches to the Assembly, 1938-39.

1939: In January, he delivered to the Gokhale Institute of Politics and Economics a lecture called "Federation versus Freedom".

During the debate over Congress's plan to leave the government in protest at not having been consulted about the declaration of war on Germany, Dr. Ambedkar made his own loyalties very clear: "Wherever there is any conflict of interests between the country and the Untouchables, so far as I am concerned, the Untouchables' interests will take precedence over the interests of the country. I am not going to support a tyrannising majority simply because it happens to speak in the name of the country.... As between the country and myself, the country will have precedence; as between the country and the Depressed Classes, the Depressed Classes will have precedence."

In November, Congress left the government. Jinnah arranged the celebration of a "Day of Deliverance," and Dr. Ambedkar enthusiastically joined him. Dr. Ambedkar was careful to emphasize, however, that this was an anti-Congress rather than an anti-Hindu move; if Congress interpreted it as anti-Hindu, the reason could only be, he said, that Congress was a Hindu body after all.

Dr. Ambedkar was now lecturing and writing constantly, and was heavily involved in politics. With Independence (and Partition), he joined Nehru's cabinet as India's first Minister of Law, and became the Chairman of the Drafting Committee for the Constitution. Framing the Constitution and guiding it through to adoption was his greatest political achievement.

1940: In December, Dr. Ambedkar published the first edition of his "Thoughts on Pakistan". In this work he argued that though partition would be an unfortunate thing, it wouldn't be the worst possible outcome, and if the Muslims wanted it they had a perfect right to claim it.

1942: He founded his second political party, the All-India Scheduled Castes Federation, which didn't do so well in the elections of 1946.

Dr. Ambedkar was inducted into the Viceroy's Executive Council as Labour Member, a position which he held until his resignation in June 1946. His thoughtful comments in that forum cover various topics.

Congress started the "Quit India" movement. Dr. Ambedkar severely criticized this move. He described it as "both irresponsible and insane, a bankruptcy of statesmanship and a measure to retrieve the Congress prestige that had gone down since the war started. It would be madness, he said, to weaken law and order at a time when the barbarians were at the gates."

1943: On January 19th he delivered the Presidential Address on the occasion of the 101st birth anniversary of Justice Mahadev Govind Ranade. It was published in book form in April, under the title "Ranade, Gandhi, and Jinnah".

In September he also prepared and published the vigorous memorandum, "Mr. Gandhi and the Emancipation of the Untouchables".

1944: On January 29th, he presided over the second meeting of the Scheduled Caste Federation, in Kanpur.

1945: In February, he published a revised version of "Thoughts on Pakistan"; this second, expanded edition was called "Pakistan; or Partition of India". A third edition of this book was published in 1946.

On May 6th he addressed the Annual Conference of the All India Scheduled Caste Federation, held at Parel, Bombay. This speech was soon published as "The Communal Deadlock and a Way to Solve It".

In June, he published a political manifesto detailing the problems of dealing with Congress, and accusing it of many acts of betrayal: "What Congress and Gandhi Have Done to the Untouchables". The next year, he published a second edition, with major revisions in one chapter.

1946: In June, he founded Siddharth College, in Bombay; it was a project of the People's Education Society, which he had founded in 1945.

In October, he published "Who Were the Sudras? How They Came to Be the Fourth Varna in the Indo-Aryan Society". He dedicated the book to the great early reformer Jotiba Phule.

1947: In March he published "States and Minorities: What Are their Rights and How to Secure them in the Constitution of Free India", a memorandum on fundamental rights, minority rights, safeguards for the Depressed Classes, and the problems of Indian states. Dr. Ambedkar accepted Nehru's invitation to become Minister of Law in the first Cabinet of independent India. On August 29th he was appointed Chairman of the Drafting Committee for the new Constitution.

1948: In the last week of February, the Draft Constitution was submitted for public discussion and debate: Constitutional discussions and debates. On April 15th, Dr. Ambedkar married Dr. Sharda Kabir (a Saraswat Brahmin) in Delhi; she adopted the name Savita. He was now diabetic and increasingly ill, and she took care of him for the rest of his life. In October, he prepared a memorandum on "Maharashtra as a Linguistic Province", expressing his views for submission to the Linguistic Provinces Commission. He published "The Untouchables: a Thesis on the Origin of Untouchability" (New Delhi: Amrit Book Company), as a sequel to his book on the Sudras. As always on this subject, he wrote with passion. In the Preface he said, "The Hindu Civilisation.... is a diabolical contrivance to suppress and enslave humanity. Its proper name would be infamy. What else can be said of a civilisation which has produced a mass of people... who are treated as an entity beyond human intercourse and whose mere touch is enough to cause pollution?"

In November, the Draft Constitution with its 315 articles and 8 schedules was formally introduced to the Constituent Assembly. Dr. Ambedkar concluded his speech: "I feel that the Constitution is workable; it is flexible and it is strong enough to hold the country together both in peace time and in war time. Indeed, if I may say so, if things go wrong under the new Constitution the reason will not be that we had a bad Constitution. What we will have to say is that Man was vile."

1949: In November, the Constituent Assembly adopted the Constitution, including Article 11, which formally abolished untouchability.

1950: Dr. Ambedkar gave several addresses about Buddhism; in May, he flew to Colombo, in Sri Lanka, to pursue further Buddhist connections.

1951: In February, he introduced in Parliament the "Hindu Code Bill" that he had drafted, which included greatly expanded rights for women; it proved very controversial, and consideration of it was postponed: on the Hindu Code Bill. In September, Dr. Ambedkar resigned from the Cabinet, embittered over the failure of Nehru and the Congress to back the Hindu Code Bill as they had earlier pledged to do. He became the "Leader of the Opposition".

1952: Dr. Ambedkar received an honorary L.L.D. degree from Columbia University as part of its Bicentennial Special Convocation. The President described him as "one of India's leading citizens—a great social reformer and a valiant upholder of human rights."

1953: His political thinking included analysis of the issue of linguistic states; he published "Need for Checks and Balances" on this question. In 1955, he was still working on the subject, as the preface to "Thoughts on Linguistic States" testified.

1954: In the midst of his round of (increasingly embittered) Parliamentary and other activity, his health gave way; he was confined to bed for two months.

While dedicating a new Buddhist vihara near Poona, Dr. Ambedkar announced that he was writing a book on Buddhism, and that as soon as it was finished, he planned to make a formal conversion to Buddhism. He also claimed that the image of Vithoba at Pandharpur was in reality an image of the Buddha, and said that he would write a thesis to prove this claim.

1956: Dr. Ambedkar brought the manuscript of "The Buddha and His Dhamma" to completion. "In February 1956 two new chapters are added to it: 'There is no god'; 'There is no soul'.... On March 15, 1956, Ambedkar wrote the Preface to his book in his own handwriting and dictated it to Rattu [his secretary]." Printing began in May, but was slowed by constant last-minute revisions of the proofs.

From June to October, he was bedridden in his Delhi residence. His eyes were failing, he suffered from side effects of the drugs he was given for his diabetes, he felt deeply depressed.

His formal conversion took place on Oct. 14th in Nagpur, a town selected for reasons he explained in his moving speech, "Why Was Nagpur Chosen?". Many thousands of Mahārs and other Dalits accepted Buddhism along with him.

In November, he flew to Kathmandu to attend the Fourth World Buddhist Conference.

On December 2, he completed the manuscript of "The Buddha or Karl Marx", his last finished work, and gave it for typing.

On the night of December 5 or the early morning of December 6, he died quietly in his sleep; on December 7 there was a huge Buddhist-style funeral procession in Bombay, and he was cremated on the seashore.

1957: "The Buddha and His Dhamma", Dr. Ambedkar's own version of a Buddhist scripture for his people, was posthumously published, by Siddharth College Publications, Bombay.

1957 and beyond: A number of unfinished typescripts and handwritten drafts were found among his notes and papers and gradually made available. Among these were "Waiting for a Visa", which probably dates from 1935-36.

Liberalism and Reformism

By upbringing and training Ambedkar was influenced by western liberalism. The openness and liberal values of the western society had struck him with pleasant surprise by his own admission. There is a reason to believe that he had studied Marxism. His first essay on caste reflects some amount of analytical orientation of Marxism. One of the subjects in his curriculum also happened to be related to Marxian socialism and his guide Prof. Seligman was well versed with the economic interpretation of history. However, as his later work reveals, Ambedkar reflected more closeness with the liberal tradition than Marxism. However, consciously he never identified himself with the Liberalism. Being aware of its pitfalls, he needed to declare that he was not a liberal reformist, although while having reservations with the postulations of Marxism he could never hide his attraction towards it.

The pitfall of his thinking emanates from his conception of the moral force of religion divorced from the material reality. He therefore hopes that without any bloodshed, the society based on liberty, equality and fraternity could be created. Of course as hypothesized above, he conceptualizes the constitutional State based on these principles. With this wishful thinking, he tends to ignore the fact that regardless of the pretensions of ruling classes, the impact of liberal governance in the multicentric iniquitous society is bound to result in sustaining multicentricity and inequality.

This liberalism rather promotes politics of casteism and communalism, schism among dalits, their use in political power games, subversion of their real problems and protects the interests of the few rich. It was a kind of contradiction in terms to assume that liberal democracy, which is actually the manifestation of the political power of the bourgeois, will do justice to the paupers. It might appear to extend certain concessions to the weaker sections, but its real motive is to maintain the existing rule of the ruling classes. Liberal democracy might appear better than the decadent Hindu caste system but it is incapable of bringing any real change in favour of dalits. It muffles the tension of the exploitative system and kills the revolutionary motivation of its victims.

Redefinition Project

Many of the constructs employed by Babasaheb Ambedkar in his working have a qualified meaning. Firstly, they are not absolute as they appear. They are the derivatives of his thought process, the source of which could be traced to his basic objective of annihilation of castes and creating a society based on equality, liberty and fraternity. Even these three principles that he held so dear to his heart, bear very different meaning from the familiar ones associated with the French Revolution. He said he had them from Buddha. What Buddha said also is to be understood from his interpretation, which could be as different from the accepted version as to be disproved by the Buddhist church.

His Buddha and His Dhamma, for instance, had faced this kind of disapproval initially from many Buddhists. Understanding Ambedkar thus essentially demands extra consideration and care about the specific meanings of the constructs and concepts he uses. The lack of it has already caused much misunderstanding among many people. It is one thing to have a clear understanding of what he said or meant but quite another to extrapolate it to something congruent to his basic objective or vision that may be useable in shaping the future movement.

Quite like Marx had said of philosophy, it could be said that the issue is not to understand Ambedkar as he is but to possibly think of him as a weapon in the struggle to which he devoted his life. The redefinition referred to here will have to essentially address both these issues. From the viewpoint of one seeking a revolutionary change, there are indeed many dimensions on which Ambedkar calls for critical interpretation. Many of the concepts that seem to act as the props for his formulations are rooted in the reactionary camp.

Paradoxically, he brings them to work for his emancipatory project, which potentially is no less than a revolution. Predominant among these concepts are identified as State, religion, liberal democracy, constitutionalism, revolution, socialism, violence and Marxism, that some way or the other have been the cause of misunderstandings about him.

It is important to appreciate that Ambedkar employed the search process that is essentially rational and the underlying objective undoubtedly radical. There could be flaws in the specific design or the application of the search process, depending upon the State of his knowledge and complexity of the situation to which it is applied. Besides this, the end result depends upon the repertoire of alternatives used for the search. What it means is that the specific method, thought or action of Ambedkar may constitute the historical facts but they cannot be taken in their face value if one wants to comprehend the ideological aspects of Ambedkar. In his usage of the above concepts for instance he does not always exercise the academic rigour.

Besides the reason that much of his usage was addressed to the un-academic lot, most of the times he tended to impart his own meanings to the terms he used. With the changed contexts or with the change in information, he readily changed his opinions. For, hypothetically speaking, if Ambedkar had lived longer he would have certainly changed his views, looking at more information available or experiencing the undesired aftermath of some of his own beliefs and opinions.

Had he not disowned the Constitution, which he had so laboriously written and so forcefully defended, saying that he was used as the hack to write it? Whatever he had done had several limitations. He never hesitated to change his opinion or stand if he was convinced that it was right. The redefinition project proposed here, in a way, is something, which he has done himself, all his life and would have continued doing if he had been alive. It is essentially something in the nature of continuing his unfinished task.

The methodological aspects of this exercise consists in the process of conceptualisation of the core vision and ideological proclivities of Ambedkar through the analytical study of his life within its contextual parameters, oriented towards capturing its intransient content.

It should reflect the basic purpose, that is, to see whether and how he could catalyse the emancipatory movement of dalits and in turn democratise the Indian society. This process may not be free from bias. The bias could be in favour of the change craved for by the have-nots, not of the ruling classes that has necessarily been colouring the history so far.

'Ambedkar' for the Movement of Dalits

'Ambedkar' for the Dalit movement, first of all, should be shorn of all the sectarian outlook that unfortunately came to be associated with him. He was an iconoclast and therefore should inspire us to break such icons that are imbued with this outlook. Dalits have to demolish all the handiwork of the reactionaries and vested interests. The project of redefinition of Ambedkar should liberate him from the dens of the ruling class and bring him back to the huts in slums and villages where he rightly belongs.

The greatest thing about Ambedkar is his consistent anti-dogmatic stance. He never accepted any thing in name of authority. He hated humbug of every kind. He always approached problems with a student's sincerity and researcher's intellectual honesty. He gave a vision that even the ideologies are bound by the tenet of impermanence and no body should claim them validity beyond their times. His followers therefore can assume absolute liberty to think through things as per their own experience in changing times.

'Ambedkar' against Exploitation

The underscoring vision in Ambedkar's thought and action is to be found in his yearning for the end of all kinds of exploitation. Whenever and wherever he smelt exploitation, he raised his voice against it. The caste system that subjugated more than one fifth of the population to levels worse than animals' for more than two millennia and which represented institutionalisation of the most heinous inequality by the Hindu religion as ordained by its gods, became the prime target of his life. He attacked it from the standpoint of its victims—the untouchables. He waged many battles; initially targeting the citadels of Brahminism—the custodian of the Hindu religious code, and later politicised the battle, realising the ineffectiveness of the former. He did not let this objective out of sight even for a moment and worked incessantly for its achievement. This Herculean task almost completely overshadows the fact that his struggles extend well beyond the caste struggles and rather encompass all other forms of exploitation.

Even the credit for struggling against untouchability was reluctantly granted to him by the establishment which had belittled him initially as merely a leader of his own caste—Mahar. This prejudicial treatment of Ambedkar could itself be taken as a measure of the severity of the problematic of caste. The facts are clear today that not only was his struggle directed towards the emancipation of all the untouchables but also towards annihilation of the entire caste system. It was basically against the systemic exploitation that ran unabated for centuries. The protest against this inhuman system could be articulated only in a concrete situation, not in a vacuum.

He did not theorise the struggle on a hypothetical plane. He built it on the basis of real problems in a concrete situation. Unlike many cases, the motive force for his life mission was provided by his experience itself. Although he pitched his tents against Brahminism, he never bore any enmity against the Brahmins or identified any one for his friend or foe by caste. The Bahishkrit Hitkarini Sabha that was the launch vehicle of his movement had majority from the forward caste people in its executive body. Even later, this intention of having a non-caste base for the organisation could be consistently seen in his movement, be it the Mahad struggle or the Indian Labour Party. He was perceptive enough to say that the Brahminism could exist in all the castes including the untouchables, for that was the essence of the casteism. It is tragic to find his legacy being monopolised today by only the scheduled castes.

Although, he considered the magnitude of the problem of emancipation of dalits is such as to warrant his sole attention, he did take cudgels for other oppressed entities like workers, peasants and women. At one occasion in response to the accusation that he did not care for the tribals, he had to squarely admit the fact that he considered the problem at hand big enough to outlast his life and provokedly put that he never claimed to fight for whole humanity. Such instances though disturbing enough could be understood within their specific context. While dealing with the socioeconomic depravation of dalits, he comprehensively exposed certain systemic dimensions that help perpetuate exploitation. For instance, he was well aware of the capitalist and imperialist oppression besides the decadent feudalism within which domain his problem lay.

Capitalism

During the colonial British regime, capitalism started taking root in India with the collaboration of Indian mercantile capital and British capital. Unlike Europe, it did not have to battle against feudalism; rather it was implanted on the trunk of the latter in India. As a result, even in the capitalist institutions in the cities, caste discrimination simultaneously existed. Ambedkar was quite aware of the exploitative potential of capital and hence he had declared capitalism and Brahminism as the twin enemy of his movement. Capitalism was in an infantile stage then but Brahminism encompassed the phases of slavery, feudalism and extended its tentacles as we see to the phase of imperialism. Moreover, he noted the reactionary compradore character of rising capitalism in the contemporary sectors of the economy and the inhuman exploitation of workers that it unleashed.

His, Indian Labour Party (ILP) was an attempt to take up the question of capitalist exploitation, as well as to combine the struggle on both caste and class basis. Various workers' problems were taken up by the ILP, the leadership of combined strike of the mill workers, parliamentary fight for the workers' interest in relation to the Industrial Disputes Act, and various legal reforms that were brought about while he was in the Executive Council of the Viceroy, can be the examples of his concern for workers' exploitation. It cannot be denied that his approach to these contemporary problems was closer to that of the Fabian socialists with whom he was more familiar. But, it was a model adopted out of familiarity and pragmatism, a matter of strategy, never thought out on an ideological plane to be a theoretical plank. Although, there cannot be any doubt that he stood against capitalism, he could not articulate a sound theoretical basis for doing so. Resultantly, his efforts remained constricted with a short view of workers' welfare but could not provide them a vision of their liberation.

Imperialism

Ambedkar's attitude towards imperialism has been projected in a distorted manner right from the beginning, mainly because he refused to take part in the freedom struggle or opposed Gandhi who for certain category of ignoramuses was the anti-imperialism personified. He strategically sought to maintain neutrality vis-a-vis the colonial State. As per him, it would not be possible for the resourceless dalits to fight its mighty foes all together. He did not want to dissipate and squander his extremely limited resources on several fronts.

He however knew the basic exploitative character of the colonial regime. At several occasions, he burst out saying that British imperialism and Indian feudalism were the two leaches that clung to Indian people. However, there was a fundamental difference between his and others' viewpoint. For instance, he did not approve equating opposition to imperialism with opposing the British. He noted that the opposition to imperialism couldn't be effective until its supporters within the country are left untouched.

The then leader of the Communist Party of India—Manabendranath Roy once met him at his residence and during discussions insisted that destruction of imperialism had to be the first and foremost objective of Indian politics. Ambedkar's response to him summarised his outlook towards this problem. He replied to Roy in explicit terms that without struggling against the landlords, mill owners, moneylenders—the friends of imperialism within the country, it was not possible to wage an effective fight against imperialism. It may be a matter of research but a priori his anti-imperialist attitude pervades even his writings as a student.

The validation of his stand comes from an entirely unrelated corner and nearly half a century later. Suniti Kumar Ghosh, (1985 and 1995) in his books has shown in great detail how the Congress representing landlords and capitalists had played a compradore role to serve the interest of imperialism during the so-called freedom struggle and how even after the transfer of power in 1947 the grip of imperialism instead of weakening became stronger. Does it not indicate that he was more correct than any of his contemporary politicians? The ones who biasedly wish to pronounce their half baked verdict that Ambedkar was a stooge of British merely on the basis of his acceptance of membership of Viceroy's Executive Council or talking to the Simon Commission, not only display their ignorance of history but also their casteist fangs. They ought to rethink the comparison between Ambedkar who, even being apparently a part of the imperialist apparatus was perhaps striking at its roots by empowering the people and many others, so called nationalists, who after wearing the mask of anti-imperialism were indirectly strengthening its pillars.

In relation to British rule, Ambedkar basically makes two points. The first is that he questions the so-called freedom struggle launched under the leadership of Congress as an anti-imperialist struggle. He contended that the Congress basically represented the class of feudal lords and the urban capitalists—the two some exploiters of Indian masses. Although, it succeeded through the charismatic leadership of Gandhi in galvanising masses in its support, it essentially relied on bargaining with the colonial rulers for securing itself more share of power.

It always throttled the mass spontaneity as in the case of 1942-uprisings and actively opposed the genuine anti-imperialist struggles of the revolutionaries like Bhagat Singh. Ambedkar reflects the understanding of true character of the Congress in his own way, when he says that if Congress was fighting a real anti-imperialist war, he would whole heartedly support it. The rhetoric of such statement apart, for he never appears to even take a note of other truly anti-imperialist struggles like the one referred to above, it is enough to reveal his attitude towards imperialism and understanding of the class character of the Congress. He knew that the class character of the Congress would not permit it to don this role in reality. Ambedkar could see through the anti-imperialist masks the real fangs of an exploiter of masses.

He thus not only saw no point in siding with this more real exploiter of people than perhaps the colonial rulers, but also did not hesitate to openly oppose it when it came in the way of Dalit liberation. He smelt rot in all such struggles that refused to notice existence of inhuman exploitation of some of their own people within their precincts and tended to over-externalise their woes. Here lay his second point when he raised a question of Hindu imperialism perpetrated through its caste system that was certainly seen as more vicious by its victims than the British rule. It may be pertinent to ask of those who raise the issue of Ambedkar's attitude and conduct towards imperialism, to answer as to why the problem of untouchability or caste system that reduced its one—fifth of the population to subhuman levels did not find a mere mention in the lofty discussions of freedom struggle that predated Ambedkar's raising it. The anti-imperialist aspiration also could be seen in the context of the class/caste division in the society.

The battle for the lost kingdom waged by the vanquished lords also could be camouflaged as an anti-imperialist struggle and at the same time the genuine peoples' anti-imperialist aspiration manifested in the form of say anti-feudal struggle could be condemned as the pro-imperialism, merely because it directed its gun towards the props of the imperialism. The real anti-imperialist aspirations belong to the masses of people the manifestation of which is possible only through the peoples' war. Whatever anti-imperialist struggle people waged were soon hijacked by the phoney war whose real intent was to extract political power to native ruling classes. While the scenes of anti-British struggles were being enacted for the 'mother' India of exploiters, Ambedkar busied himself to liberate the other India- the India of the exploited and oppressed.

Oppression of Women

Besides these mainstream forms of exploitation even the subaltern forms like women's exploitation, could not escape his agenda. He viewed them as the most oppressed of all. His approach to the problems was typically that of a liberal democrat constitutionalist. This certainly constrained his articulation of this problem as in many others. This issue will have to be seriously rethought by dalits under the redefinition project. But suffice here to say that at any opportunity, he raised his voice against women's discriminatory situation in the society.

His basic law of social engineering was that the social revolutions must always begin from the standpoint of the most oppressed or the ones on the lowest rung of the society. Right from the days of Mook Nayak and Bahiskrit Bharat, he appears to take cudgels for women. He always involved women in his struggles and tended to give them vanguard positions. For example, about 500 women had marched at the head of the historical procession at Mahad to assert the untouchables' right to drink water from the public tank. He was immensely pained to see the permanent denial of education and religious rights to women ordained in the Shastras of the Hinduism (*e.g.* Manusmriti). His democratic consciousness never reconciled with any thing lesser than the equality of men and women though its expression was acutely constrained perhaps by his anxieties about the possibilities, so much so that it might even be mistaken as the male centric tactic.

While he asked women to be good mothers so as to shape up their son or to be good wives to their husbands or be a carrier of community's cultural baggage, he did struggle for their equal rights as in the case of Hindu Code Bill. He described sacramental marriages (Mathew, 1991) as polygamy for men and perpetual slavery for women because under no circumstances within that system the latter would get liberty from their husbands, however bad or undesirable they may be. He insisted that women should have the freedom to break this contract.

On Revolution

Revolution, on the face of it, appears to be an anathema to Ambedkar who seems to dread it and instead advocates reforms. But it would be disaster to take it at face value. For, like many other terms, his usage of the term 'revolution' does not bear the same meaning as is in vogue, particularly in Marxist circles. What Ambedkar seems to detest in revolution is the violence. At many places he tends to equate rebellion, revolution and revolt with violence. He also seems to disagree with the method of insurrection. He thought that without mass consciousness being ripe enough for revolutionary change, insurrectionary methods would not succeed.

Moreover, he appears to be sceptical of the justness of revolutions as they invariably represent the triumph of the collective over the individual. It may be attributed to the influence of liberal democracy in which he got his indoctrination in his formative days. Liberal democracy always put an individual on a high pedestal and considered it precious. His concern for the individual is not again doctrinaire but emanates from the value that any and every human being is precious and the belief that alone can act as the best guarantee against the collective tyranny and totalitarian excesses any time. In the context of the collapse of erstwhile Soviet block, where the totalitarian states that came into being in the name of dictatorship of proletariat and played havoc with people, this human-centric value assumes importance. One needs to be however reminded that even the values do need to have some material bases. They do not fall from sky. The contradiction between collective and individual has thus to be resolved in the concrete situation. As for Ambedkar, apart from this scepticism, he does not seem to have any dispute with the general aim and object of revolution.

Revolution, inasmuch as it seeks to bring about a fundamental change in the social relations in the society, will always be opposed by the forces of status quo, whose material interests are directly threatened by this change. In corollary, it becomes imperative for the forces of revolution to overcome this resistance whatever be the means. Whereas, the antagonist camp will always project codes of ethics and morality for their tactical defence in face of the onslaught of revolution, the revolutionaries discard them as decadent; for them revolution itself represents the highest value. Being the upsurge of the suppressed ones, revolutions do have a tendency to be bloody, but it is always in response to the resistance of its opponents.

Therefore it is a representation only by the vested interest to associate violence or moral turpitude with revolutions. On the contrary, it would be more logical to say about the revolutionaries that being propelled by an external motivation to deliver mankind from the existing traps at the risk of their own lives; they cannot be bloodthirsty people. The revolutionary violence is almost an inevitability that arises at the instance of the oppressors.

In that sense Marx called "violence as the midwife of history", emphasising its inevitability. It is the inevitability that marks the compulsion of the vast majority of people to resist the antirevolutionary violence of the minority. The prerequisite here is that the revolution truly represents the majority consciousness. If it does not, then the violence of the few exercised with howsoever a lofty objective could transform into its antithesis, fortifying itself against the majority will. In this sense and insofar as the recent history showed the scepticism that revolutions trample upon individual's rights therefore cannot be dismissed as baseless.

Historically, revolution is the process of identifying and destroying the obstacles in the existing order to take productive forces to the qualitatively next higher level, for the overall progress of human race. The progress achieved by mankind so far is basically due to the qualitative transformation from quantitative continuum that characterises revolutions. The qualitative transformations always need concentrated inputs, akin to latent heat in the case of phase transformation of water. In social transformation it takes the form of revolutionary energy that in turn may manifest into violence. Ambedkar does not neglect the necessity of violence. As he himself said that if dalits wanted to be effective they would need the canons.

It is erroneous to construe that his opposition to violence was idealistic or doctrinaire. Violence was not a taboo for him; it could be practised when it was absolutely necessary. Even his mentor Buddha, who is respected as the greatest apostle of non-violence had the same pragmatic approach towards violence. Ambedkar was not obsessed either with the idea of non-violence or the value of individualism professed by the classical school of liberal democracy. His reservations were against the possibility of the cunning of a few overriding the will of majority as had happened in the case of caste institution. He would hate to see any thing like caste getting institutionalised again. In his scheme of things he therefore was not ready to compromise the value of democracy, the will of majority of people, whatever may be the end.

Ambedkar did not juxtapose reform against revolution, as many people tend to do. He does not reflect comprehension of technicality of dialectical materialism in his usage of these terms. Often his revolution is the violent overthrow of the existing rule and establishment of the new rule. Likewise, he does not seem to mean that reforms will not entail a qualitative transformation. What he certainly means by revolution is the change brought about in a 'big bang' manner. Notwithstanding what Ambedkar said about his own work, revolution does not always entail a 'big bang'; it is not a point concept as mistakenly regarded in common parlance, but a line concept. Mao's Great Proletarian Cultural Revolution may be a good example of this.

There could be many bits of work, which contributes to taking society to a qualitatively higher rung in the ladder of progress that qualify to be the revolutionary work. The qualitative change itself occurs over a discrete time horizon and not a moment. The moment that marks out transformation of power and looks like a 'big bang', alone is not the revolution. The particular phase of history puts constraints on the kind of changes that can be conceived in its womb. Some one dreaming of a socialist revolution in the slave society would only at best be a daydreamer; a romanticist but he cannot be a revolutionary. Likewise, certain phase of history demands a lot of quantitative preparation before a revolutionary change can be planned for. Ambedkar largely reflects these kinds of concerns while dealing with the issue of revolution. He did not see Indian situation ripe enough for any revolutionary change. Any change without resolution of the caste question, according to him, would not only be detrimental to dalits but also be an extremely short-lived.

The importance of Ambedkar's work can be gauged in relation to contemporary social situation and its transitional social context. Indian society was ridden with a peculiar brand of feudalism, the most prominent feature of which was caste. Caste had incapacitated over 15 per cent of its population and maintained them at the subhuman level. The large part of the balance population also suffered the degradation in a varying degree. This decadent institution had far outlived its minimal utility and as a result for centuries kept Indian society in a fossilised form. It served the material needs of a handful of people but all perceived varying stakes in the system on account of its hierarchical structure and faithfully practised it because they internalised it as their *Dharma* ordained by none other than God.

The possible exception in generic terms were dalits who were placed at the lowest rung of the caste ladder and had hardly anything to their share. However, in particular terms only a few castes from the Dalit castes, who did not have any specific caste profession and hence little stake in the system and consequently who as the general workmen of villages had better exposure to the changing urban life than any one else came out of the hegemony of Brahmins to articulate the challenge to the system. The hierarchy among dalits however prevented them to come together and consequently this challenge had to be articulated caste by caste. Capitalism that took root in India in big cities also had struck compromise with the caste institution like its harbingers, the British imperialism. It was an arduous task, as it still is, to conceive a model for this struggle and still more difficult to build.

During the colonial times for various reasons these struggles had germinated largely in Maharashtra and Southern states where the social structure reflected sharper polarisation between dalits and balance society. Ambedkar's advent in the Indian sociopolitical scene marked their zenith. It articulated its attack on Brahminism and capitalism that accepted its alliance, focused its organisation on dalits and gave a clarion call for annihilation of castes for achieving the ultimate aim of society based on Liberty, Equality and Fraternity to provide wider umbrella for all progressive forces to work. It reflects the distinct historical need to democratise Indian society without which it was bound to suffer constriction of its productive forces. This work had to have large content in the sociocultural realm; it is a credit to Ambedkar's acumen that he gave it a political dimension.

It had to be approached as reform. Ambedkar clearly found the talks of communist revolution as out of phase with the history although he never fully accepted the tenets of historical materialism as he thought it negatived human ingenuity and carried it through a predetermined channel. He insisted that India did not provide congenial soil for germination of class-consciousness because of castes. Their annihilation therefore constituted the first task in the revolutionary agenda. It is unfortunate that many communist revolutionaries still parrot the same characterisation of his work as in years back their predecessors proclaimed using the spoon fed theories from the West.

One day it is hoped that the contributions of all the caste struggles to democratisation of Indian society would be restored as a native revolutionary heritage by these well meaning people. Till then it will always sound puerile to pigeonhole the historical work as reform or revolution merely on the basis of syntax and not the content. However, it is much more unfortunate for Ambedkarite dalits to deny themselves the credit for this historical contributions by dissociating from the revolutionary agenda, mistakenly thinking that it something alien to them. Effectively, not only they are denying themselves a historical opportunity to contribute to revolution but also delaying their own emancipation.

Dalits have to rethink their position vis a vis revolution. Ambedkar's dream of a society based on liberty, equality and fraternity cannot be realised except through revolution. They will have to understand Ambedkar's life and mission only from this perspective. His contribution to Indian revolution lies in the fact that he tried to comprehend Indian reality independently and tried to contribute to the resolution of its contradictions in his own way. Indian history held out the gauntlet of fossilised Indian feudalism for so long to the Indian revolutionaries but every one conveniently wished it away, initially as a superstructural matter that would disappear automatically when the material base is revolutionised and now after seven decades as a problem belonging to both structure and superstructure, that could be solved through revolutionary practice. It still lacks the clarity and courage to hold the bull by horn.

Ambedkar did not confuse issues, he saw clearly that the annihilation of caste will have to be consciously worked for before taking up any revolutionary project. He went beyond and found out the institutional base of castes in the Indian village whose economic support lay in the land-relations and caste division of labour. But, unlike many communists who still use the stereotype of land reform—a slogan of land to the tiller as the only revolutionary programme, he did not hamper on it because he knew that firstly it was economically impossible to satisfy the land hunger of the landless in the country, secondly the likely transformation of landless to a marginal farmer through land reforms was unlikely to solve the problems of dalits and thirdly, as the later empirical data showed, contrary to expectations the land reforms could aggravate the problem of caste. Instead, he proposed nationalisation of land and co-operativisation of farming.

He realised the necessity of detaching substantial village population from land and absorbing it into the industrial sector that was to be mainly under State sector. Even in retrospect, these points could have constituted a viable agenda for democratic revolution. As one naxalite scholar—(Ashok Kumar, 1995) perceptively puts it, for having independently seen the question of annihilation of castes linked with the question of land one could unhesitatingly call Ambedkar as the torchbearer of the people's democratic revolution.

2

'Ambedkar' as Thinker

Dalits are never tired of projecting Ambedkar as the greatest of all the leaders. That unfortunately smacks of sectarian attitude and of their blind devotion to him. They need to understand that the measure of greatness of any person could only be her / his contribution to better the human situation, in terms of correct understanding of its ailment and contribution to cure it. What Ambedkar did could be seen in relation to the broad five currents in Indian politics of his times:

- The Reformists current that wanted to bring about development on the western pattern, possibly with the support of British imperialism,
- Congress, that represented Indian capital and which demanded self-rule under the domination of British imperialism,
- The Terrorist Nationalists who had taken up arms in their fight for freedom against British imperialism,
- The Communists who were trying to implant Bolshevik revolution in India, and
- The Muslim League which opened up a separatist front of Muslims.

All of them scarcely reflected an understanding of the Indian situation. For instance, none showed even a cursory concern about the problems of one fourth of their countrymen who were forced to live worse than animals as ordained by their decadent religion. It was indeed surprising that although all craved for self-rule from the British, none concerned with the caste-system which basically was responsible in pushing the country repeatedly into slavery. None seemed to attempt an objective analysis of either the history or the present of this country. It could circumstantially be said that their motivations came from their narrow class-caste interests. These movements were motivated by the desire of an abstract freedom for country and a refusal to see the concrete slavery of their own people. Granted that the problems before the country were really intricate, still no one would dare say that the need for democratisation was in anyway subordinate.

The real people's movement in the country was required to wage simultaneous war against imperialism, internal compradore bourgeoisie, landlords and Brahminism. It was only Ambedkar who clearly indicated this requirement. In this light, he was certainly ahead of all others. His own bitter experiences with untouchability had stood him in good stead in seeing this more clearly than any other. He strove to build his movement along this understanding but unfortunately it was neither in his power to deal comprehensively with all the issues, nor was there an ideological and programmatic clarity required therefore. He inevitably had to focus his attention on dalits who were the worst victims of this multifaceted oppression. It was the misfortune of Indian history that this struggle progressed in a constricted manner and eventually got dissolved into regressive statist politics. It reflected both the limitation of Ambedkar as well the situational compulsion on him.

The anti-caste movements before Ambedkar were mainly welfare oriented. Some wanted a higher rank for their own caste in the caste hierarchy and some taking the inferior culture of their caste to be the reason for their suffering, aimed at improving the same. Mahatma Phuley's movement was an exception to this trend insofar as it attempted to unite the Sudra and Ati-Sudra castes against the exploitation by the parasitic castes of Shetjis (capitalists) and Bhatjis (priests). While Ambedkar accepted the lineage/inheritance of this movement and held Phuley in greatest esteem as his one of the three *Gurus*, he went beyond to declare annihilation of caste to be the object of his movement in the direction of the goal of 'liberty, equality and fraternity'. In the historical context it certainly was a radical step. He rightly diagnosed that the caste system is basically sustained by the peculiar economic constitution of the Indian village of which the land relations were the main features. Towards breaking this link he toyed with an idea of separate settlement for dalits at one time and at another exhorted them to leave villages for cities.

He had clearly understood that castes stood on multiple props, *viz.*, the religio-cultural relations, feudal relations in village setting of which land relations constituted the crux and the sociopolitical nexus with the State. Annihilation of castes thus needed destruction of all of them. He soon realised the necessity of political power for this multi-fronged attack. Even to bring about the residual change in the belief system either through the cultural or religious route, he stressed the necessity of political power. In this way, for the first time he brought the problem of untouchability and caste out of the confines of culture to the political agenda.

Unfortunately, this political agenda got lost into the maze of parliamentary politics that soon became be-all-end-all with Dalit leaders. Even during Ambedkar's times the economic aspects of the problem remained largely untouched giving the impression to his followers as though they did not count. In the overall context it can be seen that they could not be as easily dealt with as the religio-cultural and political aspects of the problem. Moreover, it meant direct confrontation with the State for which Ambedkar was certainly not prepared. Alternately, the feudal relations in villages could be destroyed only if the private ownership of land is abolished and co-operativisation of farming is introduced. He thought, this structural change could be effected through the Constitution. It was a folly that he would soon realise when even as the 'chief architect' of the Constitution he failed so much as to bring this point on the agenda of the Constituent Assembly.

Conception of an Ideal

Babasaheb Ambedkar envisioned his ideal in the famous three principles: liberty, equality and fraternity. They were the basis for the ideal society of his conception. He denied that he had adopted them from the French Revolution. He said he had derived them from the teachings of Buddha. These principles were the clarion call of the French Revolution but later became the ideological props of the liberal bourgeoisie in Europe. Since Marx had ridiculed these principles as the fantasy of the bourgeois society, many people tended to stereotype Ambedkar as the petty-bourgeois liberal democrat. As according to Ambedkar the source of these principles is different from the French Revolution, familiar to Marx, there is a prima facie scope to argue that Marx's ridicule does not apply to him. His conception of these principles is indeed substantially different from that associated with the liberal bourgeoisie. Actually, what Marx refers to are the slogans of liberty and equality of the bourgeois parliamentary democracy.

There, 'liberty' is the liberty to contract and 'equality' refers to equality in market. Ambedkar insists that the conception of the ideal society ought to have them all the three together. Absence of any would not be acceptable to him. The ideal society of his dream could only be seen within a kind of spiritual frame. It would be interesting to compare this society with the communist society of Marx's conception. Marx reached his inference following the dialectical track of historical materialism. In Ambedkar' case it was just his vision. Inevitably, he had to attribute the origin of them to some spiritual source.

For Ambedkar they meant to denote the State of society sans exploitation and with an emotive ambience of fellow feeling. It was beyond him to describe this State further in concrete terms and much so to indicate the forms of struggle to reach it. Known for his obsession with pragmatism and belief that any definitive laws could not bind the flow of human history, he would avoid the speculative construction of this distant stage of human society.

Not even Marx could describe what his dream communist society would be like beyond that it would be freed of the familiar contradictions. It essentially reflected a contradiction between human desire and material reality. It would be disaster to derive the meaning of this ideal State of Ambedkar's conception from what he did. He left that to posterity to decide as per their circumstances. But, rationally there could be little doubt that the vision of Ambedkar can only be realised in the communist society of Marx's conception where most (not all) of the contradictions in human society would have been resolved. Dalits ought to internalise this vision and strive for its realisation. Ambedkar had a radical enough interpretation of his principles of liberty, equality and fraternity so as to feel inadequacy even in Marxism. He said that Marxism supported only equality. He was in need of a body of thought that would give equal importance to all these three principles. He met it with a convenient conceptualisation of religion. It is paradoxical that a person who is rational enough not to bind the posterity with his vision volitionally binds himself with what is said more than 25 centuries before.

It is natural to find ideals better articulated in spiritual spheres but it is equally true that these dream worlds are incapable to provide any clue for their realisation on the earth except for their pet prescription to ignore the material reality and imagine it happened in the mind. They run away from the fact that the evil humans suffer from are the attributes of the divisions in human society, and their abolition essentially calls for struggles by the sufferers against those who perpetrate sufferings. Howsoever, inherently rational the religion may be or radical its interpretation may be it cannot fully escape these limitations.

It can be seen in relation to Buddhism handed down by Babasaheb Ambedkar with his radical interpretation. Notwithstanding the familiar quibbling around the Dhamma and Dhamma among dalits, what remained of Buddhism with them is what would happen to any religion. It is a different question whether Marxism embodied Ambedkar's ideals or not but it is certain that they are neither realisable neither through any kind of constitutional acrobatics nor through any religious practice.

State

Ambedkar's conception of State reflects some amount of autonomy from the hegemony of the ruling classes. It is why he expected it to act as per the constitutional structure and endeavoured to incorporate the pro-Dalit bias into the Constitution. He must have realised the true nature of it, the boundaries of the autonomy and basic class bias of the State, when he actually reached not only the Constituent Assembly but also became the chairman of its Drafting Committee.

In his anxiety to secure some provisions in favour of dalits, he accepted to be the 'hack' to write what was acceptable to the ruling caste-class representatives. He must have thought that within the given constraints he had done a good job of making the Constitution responsive to the needs of the downtrodden people. Indeed, many of the provisions in the directive principles and elsewhere apparently bear clear imprint of his zeal and owe their existence to him. But, even they had to be within the strategic space provided by the rulers. His realisation of the folly was near complete when he had to burst out in utter dejection at its ineffectual implementation, that he would be the first man to burn the Constitution as it was of no good to any one. He was inaccurate, as the Constitution had proved good enough to the upper caste-class combine who had hegemonised complete political space in post-1947 India. He attributed it to the 'devils' in the Congressmen who had occupied the constitutional 'temple' he and others had built.

Ambedkar could not reach the point of understanding that the State is a mere instrument in the hands of the ruling classes to coerce the ruled ones into submission to their interests. Until the downtrodden themselves become the ruling class, they cannot expect the State to do good to them. Whatever good that appears to come to their share, in ultimate balance accrues to the other side in multiple measure.

The post-1947 State, which has never tired of propagandising its concern for dalits and poor, has in fact been singularly instrumental in aggravating the caste problem with its policies. Even the apparently progressive policies in the form of Land Ceiling Act, Green Revolution, Programme of Removal of Poverty, Reservations to Dalits in Services and Mandal Commission etc. have resulted against their professed objectives. The effect of the Land Ceiling Act, has been in creating a layer of the middle castes farmers which could be consolidated in caste terms to constitute a formidable constituency. In its new incarnation, this group that has traditionally been the immediate upper caste layer to dalits, assumed virtual custody of Brahminism in order to coerce Dalit landless labourers to serve their socioeconomic interests and suppress their assertive expression in the bud. The Green Revolution was the main instrument to introduce capitalisation in agrarian sector. It reinforced the innate hunger of the landlords and big farmers for land as this State sponsored revolution produced huge surplus for them.

It resulted in creating geographical imbalance and promoting unequal terms of trade in favour of urban areas. Its resultant impact on dalits has been far more excruciating than that of the Land Ceiling Act. The much publicised programme for Removal of Poverty has aggravated the gap between the heightened hopes and aspirations of dalits on one hand and the feelings of depravation among the poorer sections of non-dalits in the context of the special programmes especially launched for upliftment of dalits. The tension that ensued culminated in increasingly strengthening the caste—based demands and further aggravating the caste-divide.

The reservations in services for dalits, notwithstanding its benefits, have caused incalculable damage in political terms. Reservations created hope, notional stake in the system and thus dampened the alienation; those who availed of its benefit got politically emasculated and in course consciously or unconsciously served as the props of the system. The context of scarcity of jobs provided ample opportunity to reactionary forces to divide the youth along caste lines. Mandal Commission, that enthused many progressive parties and people to upheld its extension of reservation to the backward castes, has greatly contributed to strengthen the caste identities of people. Inasmuch as it empowers the backward castes, actually their richer sections, it is bound to worsen the relative standing of dalits in villages.

Thus, the State, its welfare mask notwithstanding, has viciously and consistently acted against dalits and poor people. It is a complete contra-evidence to hopes of Ambedkar who strove to maximise and make use of the autonomous space of the State for the benefits of the have-nots, particularly dalits. It is one thing to assume autonomous space but quite another to equate it with caste-class neutrality. Unfortunately, the Dalit political behaviour always reflected this erroneous notion of caste-class neutrality of the State. It has already caused great damage to the Dalit interests. The radical Ambedkar might have strategy to use State for Dalit cause but would never see it as caste-class neutral.

Socialism

Despite his ambivalence and reservations about the emphasis on the economic dimension in socialism, Ambedkar broadly remained a socialist. Some scholars do find little scope for suspecting his socialist credentials because of his disapproval of Russell's criticism of property, his non-acceptance of Marxian formulations and his placement of social issues higher than the economic and political issues. He called the complaint against love of money as 'philosophy of sour grapes' and ridiculed materialism as 'the ideology of pigs'. This impression is moreover strengthened by his reservations to accept the economic interpretation of history. But, in all fairness it may be said that what he appears to mean is the integrative consideration of all the factors that are needed for any society to be based on liberty, equality and fraternity. Notwithstanding his variant conception, there should not be any doubt about his socialistic antecedents.

His conception of socialism also underwent evolution. Once he had stated that there was hardly any difference between his socialism and communism. As such, his disagreement with the communists was about the means and not about the aim. He warns the communists that the classless society can emerge only after the emergence of a casteless society. It implies that his quarrel with the then communists was over the stages of revolution. In the 1920s and 30s, these people had borrowed the communist dogma and parroted class struggle in utter disregard of the reality.

Ambedkar, on the contrary, was firmly rooted in it. They believed that the soviet Russian model of revolution was importable for bringing about a socialist revolution into India whereas Ambedkar realistically postulated that unless the consciousness of the working class was congenial for revolution, there was no question of it materializing. And, unless the caste system is destroyed, creation of the pro-revolution consciousness was out of question. His annoyance with the Bombay communists was largely because of their dogmatic behaviour. It is unfortunate that the ideal of classlessness that was latent in his agenda never really surfaced during his lifetime. Ambedkar relies on the concept of 'State' for materialisation of his conception of socialism. His conception of State is largely idealist. He wanted the State to intervene in the economic structure and its monitoring. He wanted to constitutionalise this State intervention so that it would not be subject to change any time with the whims of simple majority vote in the legislation.

Ambedkar who taught, "The lost rights cannot be regained by making appeals or requests to the robbers; it needs struggles"; did not say anything on how the oppressed people will get such strength as to create the constitutional provisions, that would put the class structure upside down. On behalf of his party—Scheduled Caste Federation, he had submitted a draft for the future constitution to the Constituent Assembly for the independent India. It was published later as "States and Minorities". This book has really aided students in understanding some aspects of his conception of socialism.

Nevertheless, one cannot afford to forget the constraints placed by the context in which it was written. The context was that he had failed to get into the Constituent Assembly and was therefore anxious to strike a feasible and still radical note that could find the support of the vast majority of the have-nots which might then create some pressure either for its inclusion in the Constitution or for his entry into the Constituent Assembly. For some years, during the preceding turbulence of negotiations for transfer of power, he found himself totally marginalised. Notwithstanding the probable limitations of this draft, its provisions in operational terms were still very radical. The main provisions are:

- All important industries and services shall belong to the nation.
- Insurance industry shall be in public sector and insurance will be compulsory for every citizen.
- Private sector and entrepreneurs shall have a role in the economy but it shall not be dominating.
- Nationalisation of land and promotion of cooperative farming on a collective principle.

These provisions, if implemented, would have gone a long way towards supporting democratic revolution in the country. It would have limited inequality and exploitation in the economic and political sphere. Politically, it would have had far-reaching impact.

Ambedkar till the end could not completely remove the Fabian influence (which he might have gathered while in England) on him. In his times, particularly before World War II, few people in India were well versed in Marxist philosophy. The knowledge of Marxism seldom exceeded some broad principles and 'Stalin's dictatorship' painted by imperialists or the 'revolt of the workers, the insurgency of the poor. Ambedkar also does not seem to have gone very far from this point. Without indulging into the debate of ifs and buts, there should not be any iota of doubt that the ideal society of his conception could materialise only through socialism.

Democracy

Ambedkar had unshakeable faith in democracy. In his conception of an exploitation-less society, democracy has an extraordinary role. Democracy means 'one person, one vote'; and 'one vote, one value'. Democracy means empowerment of any person for participating in the process of decision-making relating to her/him, democracy means liberty, equality and fraternity—Ambedkar's definition of democracy had such a tone. Because he presided over making of the Constitution and is being projected as its chief architect, there is a misunderstanding that parliamentary democracy is what he wanted. But nothing could be farther from the truth than this. He himself spoke against parliamentary democracy. For instance, he defined parliamentary democracy as "voting by the people in favour of their owners and handing over the rights of ruling over themselves". This provides a glimpse of the expanse of his ideal, which certainly was much beyond the Indian Constitution or any common place understanding about him.

His conception of democracy appears to be purely people oriented. He showed that the bookish concepts of equality are detrimental to the disabled sections of society in the prevailing social setting and proposed a fundamental change in the concept of equality. It envisaged complete abolition of inequality. His principle of positive discrimination is based on this very concept of equality. But the operational aspects of this concept involved the need for some kind of autonomous institution, which was met by 'State' and 'religion'. It is necessary to stress that his greatness lies in the radicality of his conceptions, his vision of a human society sans any kind of exploitation; not in the remedies or apparatus he proposed in the circumstances prevailing in his time.

Aspects of the Strategy and Tactics

In the context of strategy, some of Ambedkar's contributions are really noteworthy. He brought the struggle against Brahminism into the political battlefield. He inferred that without political power the social and religious structures will collapse and motivated his followers to capture political power. But his conception of political power being acutely constrained by the parliamentary framework where bargaining is the predominant medium of securing political power, inevitably it made way for all kinds of aberrations and perversities to creep in. The prevailing politics being the game of possibilities, he was soon sucked into its vortex. Politics came to dominate the other aspects of his personality. Slowly, the impact of politics started becoming visible everywhere. This phase significantly contributed to multiplication of his inconsistencies out of tactical imperatives. Dalits have taken this legacy of parliamentary politics very seriously, almost as the be all and end all of their political being. It may be, interesting to probe how much damage this kind of political orientation inflicted on the Dalit movement.

It is altogether a different question whether Ambedkar had any other alternative to parliamentary politics for political practice. It may be argued that in the context of his resources, adversarial environment and to some extent his personal limitations, he had none. Even if it is taken as correct, it should not be forgotten that it was a matter of strategy and tactic that presuppose contextual variables: it cannot be taken as the lasting value.

The fundamental source of most of Ambedkar's political thoughts and action is his conception of State and religion, that he had adopted as the extraneous instruments to reconcile the state of flux of things and the necessity of order in them. It is necessary to understand that both, the State as well as religion, are the products of the evolutionary process of the human society. There is nothing inevitable about them. Marx took State as the instrument of coercion in the hands of ruling classes and religion as the opium of the masses. Dalits have a long experience of nearly five decades with the so-called welfare State of Ambedkar's design. What does it say? Does the State side with them in their conflict with the landlords? Does it come to their rescue when every day three or four of their daughters are raped?

Does it come to save the shame of their women when they are paraded naked in the streets? Does it take their side when they come in conflict with landlords, moneylenders in villages or with management in the modern settings? Does it really ensure they get their dues as provided in the Constitution or punish the defaulting management for noncompliance? To all of such questions the answer could only be in the negative. It is not a matter of 'devils occupying the temples' as Ambedkar lamented seeing the people occupying the Constitutional positions. The State possesses the characteristics of its master class. In India the upper caste capitalists, landlords, top bureaucrats, etc. being the ruling class, the State can never have saints who would favour dalits. It is a fact that even the government by a Dalit party like the BSP could not transform it into a Dalit State.

Religion, in Ambedkar's conception is necessary to maintain the moral order of the society. It may be interesting to examine to what extent it conforms to this idealistic expectation. Buddhism, which undoubtedly is the most rational of all the religions, does not have any evidence of having created over its near millennium long tenure in India such an order of the society which could claim liberty, equality and fraternity of Ambedkar's conception. Ambedkar's argument that the Bhikku Sangha in Buddhism was the prototype of this kind of society becomes invalid once we call into question the role of Sangha in the production process.

Before that, the order that envisaged nostalgically to recreate the value of the vanishing *Gana Rajya* in microcosm against the evil of rising monarchies, cannot just sublimate to the era of capitalism without being an alley of exploiters unless it transforms itself in some way as a catalyst for the revolution. The moral tenets of Buddhism likewise cannot submerge the lure of surplus extraction, that is its dominating ethos. If at all, it may help it become more pronounced by weakening the resistance of the exploited masses. As we empirically see, the Buddhism of his conception could not produce even a trace of morality in its adherents. On the contrary, for the masses it presents the distorted, if not inverted, worldview; it orients them to look inwards for their misery and be blind to the reality that some one exploits him. He could certainly derive mental peace and pleasure but it is a state of an intoxicated mind. Keeping in mind the causal links in Ambedkar's adoption of these instruments, the redefinition project needs to address these issues afresh.

The strategy of the ruling class always stresses on diversion of peoples' attention from their real problems and their disarmament—both in physical and ideological terms. They invariably have a multilayered strategy in place for this purpose. The vanguard parties of the ruling classes open up sentimental fronts and attempt to divert the attention of people as their real problems get aggravated. Secondly, their time-tested methods of adulterating radical ideas with the masses are always in operation. Towards this end, we see all the ruling parties vying with each other in co-opting Ambedkar. 'Ambedkar' represents a potentially dangerous ideological weapon in the hands of the Indian proletariat and so the ruling class will be hell bent on blunting its edge. They will do everything to eulogise him not only for wooing dalits for their immediate electoral gains but also to neutralise him as the radical ideological force by propagandising a distorted version of the latter. Dalits will have to exercise vigil over their ideological assets even after redefining Ambedkar.

Conclusion

For at least over last four decades Dalits have devotedly followed Ambedkar as their ideal, as a virtual God and zealously practised, as they claim, his teachings. Their social being could be seen to be totally imbued with what they call Ambedkarism—the veritable science of their emancipation. If it is true, and no one would deny it is not, it should be pertinent to ask why despite this flawless following of Ambedkar they continue to be in a pathetic state in every sense. Barring a handful dalits in the government and public sector services and of course politicians, they continue to occupy the lowest rung in the social as well as economic hierarchy of Indian society. Their politics, notwithstanding the media hype, gauged by the measure of general empowerment, continues to be in shambles. Over these decades, their relative situation either shows stagnation or decline. The insinuation to relate this State with their faith in Ambedkar itself would be distasteful for many dalits. But, it is vital for ones that are committed to their liberation to squarely face the facts and dispassionately find out where the rot lay.

Generally, beyond the first burst of anger in reaction to this question, one would face the defensive arguments to discount this relationship. Typically, they tend to attribute their miserable situation to the lack of competent leadership after Babasaheb Ambedkar, to the educated dalits who they think have become Dalit Brahmins and have deserted the community; and sometimes to the people themselves for their extreme self-centredness. Sometimes, the finger is raised at the high caste hegemony that has neutralised the impact of the Constitution that Ambedkar created. Some, particularly the leadership, even would dismiss the basic premise itself that there is something wrong with the Dalit movement. They might even go so far as to claim net achievements indicating the prosperity of themselves or some others of their like. More sober may argue that what is seen is the transition state. Conceding them all some amount of validity even would not erase the stark fact that the general situation of the vast majority of Dalit masses remains still alarmingly pathetic. Externalisation of the reasons for this state has not helped dalits wee bit. The time has come for dalits to self-critically see whether anything was wrong with their ideology and / or with their practice thereof.

Reviewing the post-Ambedkar Dalit movement at some significant milestones, one finds a queer underscoring behaviour, believed to be in accordance with the teachings of Babasaheb Ambedkar, that is certainly incongruent with the essence of what he taught. This dichotomy between the essential Ambedkar and the 'Ambedkar' in the faith of Dalit masses—the icon of Ambedkar, comes out as the problematic in this review. In relation to almost every aspect of his teaching there emerged an icon that represented varying amount of distortion. Insofar as they constituted the ideology that gripped the masses, these icons can be seen to be at the root of the Dalit pathos. Among the myriad sources for these icons, Ambedkar himself might come out as the major source. Because, even a myth can not sustain for long without some material base. The icons of Ambedkar thus could be linked to some such bases, howsoever tenuous, within his own life. The rationalist in Ambedkar never hesitated to change his opinions and behaviour if the facts so warranted. They appear to be inconsistencies to the ones who see it sans context.

His distance from the focus of control in his sociopolitical environment, the vastly varying target audience (ranging from the Englishmen to the illiterate dalits) to whom he had to communicate, the exigency to respond to the dynamics of communities set in motion by the pre-independence politics, his anxiety to accomplish as many gains for dalits as he could in his life time, the exigencies for 'short' actions at the expense of 'long' vision inevitably led to mark patterns that could support array of behaviours. The first to take advantage of it were his own lieutenants for serving their personal ambitions that set the trends of distortion of Ambedkar in the minds of gullible Dalit masses.

The ruling classes that always look for the grounds to divide masses had severally reinforced this distortion and accelerated fragmentation of dalits in every field. They, along with willing collaboration of Dalit politicians and emerging elite, promoted and sustained the particular icons of Ambedkar that would prevent political coalescence of dalits and suck them into the vortex of parliamentary politics in order to bring the establishment much needed legitimacy. Dalits failed to note this cunning and let themselves flow in the currents of confusion the ruling classes deliberately created. Today, with increasing political crises the ruling classes seem vying with each other in co-opting Ambedkar, as though that may be their last chance of survival. If dalits stood their grounds well, that might have proved to be their last act.

But, unfortunately dalits are giving them new leases of life by blinding themselves to the reality. Not only that they have not resisted these ruling class machinations but on the contrary they are curiously seen to swell the cadres of these castiest, communalist and anti-poor groups and parties. These parties who openly profess the ideology of Hindu revivalism and represent all that is decadent in Indian tradition show the temerity to project Ambedkar among their ideologies. Apart from the reasons of security that propel Dalit youth into their fold, the compradore behaviour of Dalit elite certainly has influenced the phenomenon. What is common in all the attempts by the ruling classes is to sap 'Ambedkar' of its rebellious content.

It is advantageous for them to show that he was the chief architect of the Indian Constitution, committed to parliamentary democracy and opposed to any ideology that propagates violence or revolution. He is projected to favour gradual change implying that dalits should patiently wait and strive persuasively to better their own lives. He is projected as the Bodhisatva that inspires *nirvana*—the State of total detachment from worldly matters. All these images have caused significant damage to the emancipatory struggles of dalits.

Some of these images might be the purposeful and blatant disfiguring of Ambedkar but some of them represent genuine dilemma arising from Ambedkar's own stands on various issues. From the viewpoint of the comprehensive pro-people change in the present historical phase a democratic revolution is an imperative. The motive force for this revolution ought to sprout from dalits. The history provides a strong testimony that any radical movement in the country could be sustained by dalits and tribals at its base. Ambedkar as a symbol for Dalit aspirations holds a key to the barrage that has so far bound the revolutionary upsurge in India below the alarming levels. If one concedes that Ambedkar's framework is going to haunt revolutionary commitment till the Indian Democratic Revolution actually happens and that Ambedkar represents the ideological weaponry in the hands of dalits who along with other oppressed people are going to be the axis of this revolution, then one would clearly see the need to redefine Ambedkar in radical terms commensurate with this purpose.

Many aspects of confusion with regard to the facts could be cleared with rational interpretation of his thoughts. But much might need the logical extrapolation of his basic thinking which for some reason appears to have settled in erroneous forms. As he himself showed the way in the case of his redefinition of Buddha's Dhamma, dalits will have to undertake this task, for giving themselves a powerful ideology. Their future as a social group almost hinges on this task. There are enough clues left behind by Ambedkar himself that point to this need.

There is no doubt that he was frustrated at the end of his life seeing the undesired aftermath of his lifelong struggle. He had to lament over the betrayal of the educated dalits in whom he had seen the crusaders of his mission. He had to weep with remorse that he could not do anything for his people in the villages. He had to disown the Constitution for working on which he had cut short his life at least by a few years. He had to swallow the frustration of not being able to pilot the Constitution of his conception (States and Minorities). He had to regret the anti-people State that emerged in republican India. He obviously lacked the analytical tools to see through the reasons for these happenings. His excessive religiosity and spirituality at the fag end of his life perhaps could be taken as the manifestation of this frustration.

The social engineer could only be busy with problems; he is unlikely to come to grip with the design defects in the system. Almost every thing that Ambedkar pinned his hopes on can be found today in antithetical shambles. His educational society, his vision of Buddhism, the political party of his conception, the social reforms could be some of the examples. These tragic aftermaths also would denote the necessity of a critical review of Ambedkar's thoughts if they were to be used as the ideology to further the Dalit movement towards its logical end. If this process is sincerely followed, there cannot be any doubt that this 'redefined Ambedkar' would be a revolutionary icon, organically linking the Dalit struggle to the revolutionary struggle in the world. It will truly globalize the Dalit struggle.

Political Career

Between 1941 and 1945, he published a large number of highly controversial books and pamphlets, including *Thoughts on Pakistan*, in which he criticized the Muslim League's demand for a separate Muslims state of Pakistan. With *What Congress and Gandhi have Done to the Untouchables*, Ambedkar intensified his attacks on Gandhi and the Congress, charging them with hypocrisy. In his work *Who were the Sudras?*, Ambedkar attempted to explain the formation of the Sudras *i.e.* the lowest caste in hierarchy of Hindu caste system. He also emphasised that how Sudras are separate from Untouchables. Ambedkar oversaw the transition of his political party into the All India Scheduled Castes Federation, although it performed poorly in the elections held in 1946 for the Constituent Assembly of India. In writing a sequel to *Who were the Sudras?* in 1948, Ambedkar lambasted Hinduism in the *The Untouchables: a Thesis on the Origins of Untouchability*:

> *Ambedkar was also critical of Islam and its practices in South Asia. While justifying the Partition of India, he condemned practices of Child-Marriage in Muslim society, as well as the mistreatment of women. He also condemned the Caste practices carried out by Muslims in South Asia. He was also critical of slavery in Muslim communities. He said:*
>
> *"No words can adequately express the great and many evils of polygamy and concubinage, and especially as a source of misery to a Muslim woman." "Take the caste system. Everybody infers that Islam must be free from slavery and caste. While slavery existed, much of its support was derived from Islam and Islamic countries. While the prescriptions by the Prophet regarding the just and humane treatment of slaves contained in the Koran are praiseworthy, there is nothing whatever in Islam that lends support to the abolition of this curse. But if slavery has gone, caste among Musalmans has remained."*

He wrote that Muslim Society is "even more full of social evils than Hindu Society is" and criticized Muslims for sugar-coating their sectarian Caste System with euphemisms like "brotherhood". He also criticized the discrimination against the Arzal classes among Muslims who were regarded as "degraded", as well as the oppression of women in Muslim society through the oppressive *purdah* system. He alleged that while Purdah was also practised among Hindus, only in Muslims was it sanctioned by religion. He criticized their fanaticism to Islam on the grounds that their literalist interpretations of Islamic doctrine made their society very rigid and impermeable to change. He further wrote that Indian Muslims have failed to reform their society unlike Muslims in other countries like Turkey.

In a "communal malaise," both groups [Hindus and Muslims] ignore the urgent claims of social justice. While he was extremely critical of Muhammad Ali Jinnah and the communally divisive strategies of the Muslim League, he argued that Hindus and Muslims should segregate and the State of Pakistan be formed, as ethnic nationalism within the same country would only lead to more violence. He cited precedent with historical events like the dissolution of the Ottoman Empire and Czechoslovakia to bolster his views regarding the Hindu-Muslim communal divide.

However, he questioned as to whether the need for Pakistan was sufficient and it might be possible to resolve Hindu-Muslim differences in a less drastic way. He wrote that Pakistan must "justify its existence" accordingly. Since other countries such as Canada also have had communal issues with the French and English and have lived together, it may not be impossible for Hindus and Muslims to live together.

He warned that the actual implementation of a two state solution would be extremely problematic with massive population transfers and border disputes. This claim was almost prophetic in its realization with the violent Partition of India after independence.

Architect of India's constitution despite his increasing unpopularity, controversial views and intense criticism of Gandhi and the Congress, Ambedkar was by reputation an exemplary jurist and scholar. Upon India's independence on August 15, 1947, the new Congress-led government invited Ambedkar to serve as the nation's first law minister, which he accepted. On August 29, Ambedkar was appointed chairman of the constitution drafting committee, charged by the Assembly to write free India's constitution. Ambedkar won great praise from his colleagues and contemporary observers for his drafting work. The text prepared by Ambedkar provided constitutional guarantees and protections for a wide range of civil liberties for individual citizens, including freedom of religion, the abolition of untouchability and the outlawing of all forms of discrimination. Ambedkar argued for extensive economic and social rights for women, and also won the Assembly's support for introducing a system of reservations of jobs in the civil services, schools and colleges for members of scheduled castes and scheduled tribes, a system akin to affirmative action.

India's lawmakers hoped to eradicate the socioeconomic inequalities and lack of opportunities for India's depressed classes through this measure, which had been originally envisioned as temporary on a need basis. The constitution was adopted on November 26, 1949 by the Constituent Assembly. Speaking after the completion of his work, Ambedkar said:

> *Ambedkar resigned from the cabinet in 1951 following the stalling in parliament of his draft of the Hindu Code Bill, which sought to expound gender equality in the laws of inheritance, marriage and the economy. Although supported by Prime Minister Nehru, the cabinet and many other Congress leaders, it received criticism from a large number of parliament members. Ambedkar independently contested an election in 1952 to the lower house of parliament, the Lok Sabha but was defeated. He was appointed to the upper house of parliament, the Rajya Sabha in March 1952 and would remain its member until his death.*

Ambedkar, Gandhi and Congress

Ambedkar was a fierce critic of Mahatma Gandhi and the Indian National Congress. He was also criticized by his contemporaries and modern scholars for his opposition to Mahatma Gandhi, who had been one of the first Indian leaders to call for the abolition of untouchability and discrimination.

Gandhi had a more positive, arguably romanticised view of traditional village life in India, whereas Ambedkar had a much more negative view. Ambedkar encouraged his followers to leave their home villages, move to the cities and get an education.

Limitations, Criticism and legacy: Most of contemporary sociopolitical leaders across political spectrum joined in condemnation evil social practice of untouchability, many times their priorities on the ground in eliminating these evil practices did not match to expectation of Ambedkar. While it took long time, curse of untouchability and forms of overt discrimination got removed eventually from most of Indian social life over the years.

Structure of Indian Caste system specially so of Hindu's, and their superstitions were so complex and intermingled, that those religions which in principal were not supposed to have caste system, in many instances were practising the same indirectly. While in sharing of food together by different caste people became more common, practice of marriage among same caste and political voting lines based on caste achieved very little change over the years.

The whole situation was so frustrating that while progressive people agreed with his cause, he could not get unanimous and unequivocal support from even all depressed classes at one time. So he could not get enough mileage of vote bank in actual number of parliamentary seats, in post independence era.

Benefits of Governments affirmative action did not reach adequately to affected population due to ever growing population, practices of child labour and child marriage, discrete forms of discrimination, castism among even oppressed classes themselves.

At the same time, some economically poor sections of society got disenchanted because of affirmative action as their castes or communities were not among selected few.

Above factors have been polarising Indian politics, at times, with extreme POVs. This aspect is more of reason in accusing Ambedkar as a controversial and polarizing figure in Indian politics even after his death.

While Ambedkar's supporters argue that he was working to secure Dalit and Backward Caste political rights. Contemporary and modern scholars also questioned Ambedkar's research and point of view regarding origin of the caste system and racial theories.

He acquired sympathy and criticism both about his mass conversion of Buddhism as a political stunt, from his opponents.

Ambedkar was also criticised for his intensely anti-Hindu views, though his supporters argue that he was only opposed to "Orthodox Brahminism" rather than to all Hindus. He came in touch with many progressive people belonging to Brahmin and other upper classes.

Because of oppressive traditional caste system many scholars including that of affected castes took a view that Britishers are more even handed in respect of Indian caste system, and continuance of British rule can help in eradicating some of evil practices. This political thought believed that let the social improvement happen first and then only we should go for political independence, and was shared by quite a number of social activists including that of Mahatma Jyotiba Phule and Ambedkar. So this earned criticism from certain quarters like writer Arun Shourie, who questioned Ambedkar's contributions to the Indian Independence struggle.

His criticism about certain aspects of Islam and Islamic society in India and his favour in form of benefit of affirmative action to limited sections; limited support among many Muslims.

Legacy: Ambedkar's legacy has been long-lasting on modern India. He is widely regarded as the "father of the Indian constitution" for his role in creating the document. His political philosophy has given rise to a large number of Dalit political parties, publications and workers unions that remain active across India, especially in Maharashtra. His promotion of the Indian Buddhist Movement has rejuvenated interest in Buddhist philosophy in many parts of India. Mass conversion ceremonies have been organized by Dalit activists in modern times, emulating Ambedkar's Nagpur ceremony of 1956.

Cost of Change: However, frequent violent clashes between Dalit groups and orthodox Hindus have occurred over the years. When in 1994 a garland of shoes was hung around a statue of Ambedkar in Mumbai, sectarian violence and strikes paralysed the city for over a week. When the following year similar disturbances occurred, a statue of Ambedkar was destroyed. In addition, some Dalits who converted to Buddhism have rioted against Hindus and desecrated Hindu temples, often incited into doing so by anti-Hindu elements and replacing deities with pictures of Ambedkar.

Humans as the Homo sapience evolved on this planet about 150,000 years before present. Some of the oldest civilizations known in the history of humanity have been dated 10,000 years old. Human Being is a social animal. Social animal has a tendency to govern and to be governed by a set of rules framed by the society itself. There are two fundamental types of human nature. Creative and possessive. Creative humans use human intellect for creative endeavours which enriches human thought; knowledge and wealth thereby contribute to the development of human heritage for the posterity. Possessive people, on the other hand do not believe in the use of human intellect for creative purpose. Rather, they believe in appropriation, amassing and even usurpation of the products of the labour of the creative people.

This type of people posses a strong urge to become the governing class by all means in order to achieve their aims. Lesser the degree of civilization in the society, greater is the probability of succeeding this type of people in becoming the governing class. However, in a more civilized society the creative people can offer resistance to possessive people and try to safeguard their interests. This is a continuous process in the human society. Karl Marx has scientifically analysed this conflict by applying the principles of dialectical materialism to the sphere of social phenomenon and described it as the historical materialism. Slavery, apartheid, gender bias and caste system are the abominable creations of possessive peoples for the exploitation of creative people. These are man made evils created by man for the exploitation of man. Those, who have raised their voices against these evils and given a relentless fight against the prevailing social order of their times in order to free the creative peoples from the shackles imposed on them have become immortal personalities in the human history.

Some of these great persons are better known as founders of religions. Gautama Buddha, Jesus Christ and Guru Nanak for example. Some have become famous as saints as Kabirjee, Ravidasjee and Tukarama. Some have become source of inspiration and guidance to the underprivileged classes as Krantiba Jotiba Phoolay and Periyar Ramaswamy Naicker and some are revered even more than gods as Bharatratna Dr. Bhimrao Ramjee Ambedkar. Gautama Buddha, Jesus Christ,

Guru Nanak Kabir, Ravidas, Tukarama, Krantiba Jotirao Phoolay, Periyar and Dr. Babasaheb Ambedkar they all belong to the great class of exalted Homo sapience called as Humanists.

Dr. Babasaheb Ambedkar was truly a multifaceted personality. A veritable emancipator of Dalits, a great national leader and patriot, a great author, a great educationist, a great political philosopher, a great religious guide and above all a great humanist without any parallel among his contemporaries. All these facets of Dr. Baba Saheb Ambedkar's personality had strong humanistic underpinnings. It is only regrettable that the press in the past as well as the contemporary has projected Ambedkar mainly as a great social rebel and a bitter critic of the Hindu religion. Critics of Dr. Ambedkar have ignored his basic humanistic instincts and strong humanitarian convictions behind his every act or speech through out his life. It is important to trace the origin and consolidation of his humanistic convictions.

Origins of Dr. Ambedkar's Humanistic Convictions: Dr. Ambedkar's father, Subhedar Ramji was a known follower of the Kabirpanth. Many of the Kabir's Dohas are the veritable gems of rationalism and the most daring expressions of the humanitarian beliefs. Dr. Ambedkar's mind was thus deeply imbued with Kabir's philosophy in the childhood days. On passing his matriculation examination, he was felicitated by his teacher and was presented with a copy of a book on the life of Buddha. This gift must have made a profound impact on the mind of young Ambedkar. Dr. Ambedkar stayed in America, the land of liberty, for his higher studies. There he studied the western liberal thought and the humanitarian philosophy expounded by great thinkers such as Prof. John Dewey, who was also his teacher, John Stuart Mill, Edmund Burke, and Prof. Harold Laski to name a few. The impact of this original thinker on Dr. Ambedkar's mind is evident from the frequent quotations one comes across in his writings and speeches. The contrast between the social milieu which he lived in, and the liberal academic thought he studied could not have resulted in anything but making him an ardent humanist.

Fundamental rights assured to all citizens of our country is a great leap towards establishing the basic human values in the society that was based on graded inequality. As the chairperson of the constitution drafting committee Dr. Ambedkar was instrumental in the incorporation of the principle of fundamental rights in the constitution.

Dr. Ambedkar was a firm believer in the parliamentary democracy. That is why when the fear of fascism represented by Hitler was looming large over the world,

he decided to cooperate with the British government in its fight against the fascism. Because as a humanist he could foresee the dangerous consequences of the victory of the fascism. Today some myopic people criticize Dr. Ambedkar for this. However, by criticizing Dr. Ambedkar on this score, they inadvertently expose their fascist leanings.

A few months before his Mahaparinirvana he embraced Buddhism. It was a great tribute of a great humanist to the greatest humanistic philosophy of Buddha. By initiating millions of his follower in to the Buddhist fold, he asserted his faith in the humanistic values preached by Buddha in alleviating the sufferings of his lot. He thus reached the pinnacle of the humanism by becoming a Bodhisattva.

Ambedkar and his patrons were dealt a humiliating blow by the elections of 1937. There were a total of 1,585 seats in the 11 assemblies in 'British India'. Of these 777 were 'tied'—in the sense that they were to be filled by communal or special representation from Chambers of Commerce, plantations, labour etc. Of the 808 'general' seats, the Congress, which Ambedkar, Jinnah and others denounced from the house tops, won 456. It secured absolute majorities in 5 assemblies—those of Madras, United Provinces, Central Provinces, Bihar and Orissa. And was the largest single party in 4 others—Bombay, Bengal, Assam and the NWFP.

From the point of view of Ambedkar and the British—who had been holding him up to counter the Congress claim that it represented the harijans as much as any other section of Indian society—worse was the fact that the Congress did extremely well in the seats which had been reserved for harijans. Thirty seats were reserved for harijans in Madras Presidency, the Congress contested 26 and won 26. In Bihar there were 24 reserved seats—in 9 of these Congress candidates were returned unopposed; of the remaining 15 reserved seats, it contested 14, and won 14. In Bombay of the 15 reserved seats, it secured 1 unopposed, contested 8 and won 5. In the United Provinces there were 20 reserved seats; two of its candidates were returned unopposed; it contested 17 seats and won 16. In Bengal of the 30 reserved seats, it contested 17 and won 6. In the Central Provinces of the 19 reserved seats, it contested 9 and won 5.

The lesson was there for all to see. Reporting to the Viceroy on the result in the Bombay Presidency, the Governor, Lord Brabourne wrote, "Dr. Ambedkar's boast of winning, not only 15 seats which are reserved for the harijans, but also a good many more—looks like being completely falsified, as I feared it would be."

The electorate, including the harijans, may have punctured his claims but there was always the possibility of reviving one's fortunes through politicking and manoeuvres. Efforts of all these elements were focused on the objective of installing non-Congress ministries in Bombay and wherever else this was a possibility. Brabourne reported to the viceroy that Jamnadas Mehta, the finance minister "who is chief minister in all but name", was telling him that the ministry in Bombay would survive motions on the budget and may even get through the motion of no-confidence:

"His calculations are based on the fact that he expects to get the support of the bulk of the Muhammadans, the whole of Ambedkar's Scheduled Castes Party, and of half a dozen or so of those individuals who stood as Congressmen merely to get elected," he reported. But added, "I gather that he is in touch with Ambedkar, who is carrying on negotiations for him, but, as you will find from the next succeeding paragraph, it rather looks to me as if Ambedkar is playing a thoroughly double game, in which case Jamnadas Mehta's hopes are likely to be rudely shattered."

The governor went on to report that he had also had a long conversation with Jinnah, and that Jinnah had told him that, in the event of the ministry being defeated, the Muslim League would be prepared to form a ministry provided they could secure a majority of even two or three in the assembly. "He (*that is, Jinnah*) went on to say that Ambedkar and his party were prepared to back him in this," Brabourne reported, "and that he expected to get the support of ten or a dozen of the so-called Congress MLAs mentioned above.

He made it quite clear to me that they would not support the present ministry. The governor was sceptical about the claims and assurances of all of them. He wrote, "It is, of course, quite impossible to rely on anything that Jinnah tells me, and the only thing for me to do is to listen and keep silent. I obviously cannot tell Jamnadas Mehta what Jinnah told me, or vice versa, as both of them are hopelessly indiscreet. The only thing that is clear is that a vast amount of intrigue is going on behind the scenes, but, in the long run, I cannot see anything coming out of it at all, as none of these people trust each other round the corner. Were to hazard a guess, it would still be that the present ministry will be defeated on the budget proposals and the alternative will then lie between Congress or Section 93"—the equivalent of our present-day governor's rule.

Congress ministries were formed. And in 1939 they resigned in view of the British government's refusal to state what it intended to do about Indian Independence after the War. Jinnah announced that the Muslim League would celebrate the resignations as 'Deliverance Day.' Guess who was at his side in these 'celebrations' addressing meetings from the same platforms? Ambedkar, of course. Nationalist leaders were neither surprised that Ambedkar was on the platforms with Jinnah, nor had they any doubts about the inspiration behind these celebrations. Addressing the Congress Legislature Party in Bombay on 27 December, 1937, Sardar Patel noted, "We cannot forget how Sir Samuel Hoare set the Muslims against the Hindus when the unity conference was held at Allahabad. The British statesmen in order to win the sympathy of the world, now go on repeating that they are willing to give freedom to India, were India united.

The 'Day of Deliverance' was evidently calculated to make the world and particularly the British public believe that India was not united and that Hindus and Muslims were against each other. But when several sections of Muslims were found to oppose the 'Day of Deliverance', the proposed anti-Hindu demonstrations were converted into a Jinnah-Ambedkar-Byramji protest against the Congress ministries and the Congress high command.

That rout in the election remained a thorn in the heart of Ambedkar for long. A large part of *What Congress and Gandhi Have Done to the Untouchables* which Ambedkar published in 1945 is a tortuous effort to explain that actually the Congress had not done well in the election, that in fact, while groups such as his which had opposed Congress had been mauled even in reserved constituencies, they had triumphed, and the Congress, in spite of the seats having gone to it, had actually been dealt a drubbing!

Though this is his central thesis, Ambedkar gives reasons upon reasons to explain why he and his kind have lost and why the Congress has won! One of the reasons he says is that the people in general believe that the Congress is fighting for the freedom of the country. This fight for freedom, Ambedkar says, "has been carried on mostly by Hindus." It is only once that the Mussalmans took part in it and that was during the short-lived Khilafat agitation. They soon got out of it, he says. The other communities, particularly the untouchables, never took part in it.

A few stray individuals may have joined it—and they did so, Ambedkar declares, for personal gain. But the community as such has stood out. This is particularly noticeable in the last campaign of the "Fight For Freedom", which followed the 'Quit India Resolution' passed by the Congress in August 1942, Ambedkar says. And this too has not been just an oversight, in Ambedkar's reckoning it was a considered boycott. The Untouchables have stayed out of the Freedom Movement for good and strong reasons, he says again and again.

Traditionally, according to the Hindu code of conduct, the untouchables were placed at the bottom of the caste hierarchy and were known by different names in different parts of the country. They were called *Sudras*, *Atisudras*, *Chandalas*, *Antyajas*, *Pariahs*, *Dheds*, *Panchamas, Avarnas*, *Namasudras*, *Asprusthas*, etc.

The hierarchical and inegalitarian structure of Indian society came into existence during the period of *manusmriti*. The manusmriti set the tenor of social discrimination based on birth. This, in turn led to economic degradation and political isolation of the untouchables now popularly known as Dalits. Dalits are the poor, neglected and downtrodden lot. Their social disabilities were specific, severe and numerous. Their touch, shadow or even voices were considered by the caste Hindus to be polluting. They were not allowed to keep certain domestic animals, use certain metals for ornaments, eat a particular type of food, use a particular type of footwear, wear a particular type of dress and were forced to live in the outskirts of the villages towards which the wind blew and dirt flowed. Their houses were dirty, dingy and unhygienic where poverty and squalor loomed large. They were denied the use of public wells. The doors of the Hindu temples were closed for them and their children were not allowed into the schools attended by the children of caste Hindu. Barbers and washer men refused their services to them. Public services were closed to them. They followed menial hereditary occupations such as those of street sweepers, scavengers, shoe makers and carcasses removers.

Generally the term *Dalit* includes those who are designated in administrative parlance as Scheduled Castes, Scheduled Tribes and other backward classes. However, in common political discourse, the term Dalit is so far mainly referred to Scheduled Castes. The term Scheduled Caste was used for the first time by the British officials in Government of India Act, 1935. Prior to this, the untouchable castes were known as depressed classes. Mahatma Gandhi gave them the name *Harijans* meaning children of God. Gandhi himself did not coin the name. He borrowed the name from a Bhakti movement saint of the 17th century Narsinh Mehta. The name *Harijan* became popular during 1931 amid conflicts between Gandhi and Ambedkar on the issue of guarantying communal political representation to the dalits. Gandhi took this move as a step towards the disintegration of Hindu society. By terming the untouchables as *Harijans,* Gandhi tried to persuade caste Hindus to shed their prejudices against the *achchutas i.e.* untouchables.

The purpose to adopt this new nomenclature of *Harijan* for the untouchables was to induce change in the heart and behaviour of the Hindus towards untouchables. At the same time, it was hoped that this new name would be accepted by the untouchables who would too try to cultivate the virtues which it connotes. To quote Gandhi "...probably, Antyaja brethren would lovingly accept that name and try to cultivate the virtues which it connotes... may the Antyaja become Harijan both in name and nature" (Gandhi 1971: 244-5). The term *Harijan* got further recognition as an emancipatory nomenclature in the formation of *Harijan Sewak Sangh*, an organisation established for the purpose of upliftment of the dalits under the aegis of the Congress. A weekly '*Harijan*' was also started by Gandhi to provide voice for the cause of the downtrodden. However, Ambedkar did not find any substance in the change of name for the redressal of the structural hindrances that stood menacingly in the way of the their all around amelioration. To him it did not make any difference whether the downtrodden were called *achchuta* or *Harijan*, 'as the new nomenclature did not change their status in the social order'.

The term *Dalit* was used by no less a person than Ambedkar in his fortnightly called *Bahishkrit Bharat*. Though Ambedkar did not popularise the word Dalit for untouchables, his thoughts and actions have contributed to its growth and popularity. The word Dalit is a common usage in Marathi, Hindi, Gujarati and many other Indian languages, denoting the poor and oppressed persons. It also refers to those who have been broken, ground down by those above them in a deliberate way. "It includes all the oppressed and exploited sections of society. It does not confine itself merely to economic exploitation in terms of appropriation of surplus. It also relates to suppression of culture – way of life and value system – and, more importantly, the denial of dignity. It has essentially emerged as a political category. For some, it connotes an ideology for fundamental change in the social structure and relationships". The word Dalit indicates struggle for an egalitarian order and provides the concept of pride to the politically active dalits. The word Dalit gained currency through the writings of Marathi writers in the early 1970s. "Dalit writers who have popularised the word have expressed their notion of Dalit identity in their essays, poems, dramas, autobiographies, novels and short stories. They have reconstructed their past and their view of the present. They have expressed their anger, protest and aspiration".

"Dalit" is a by-product of the Ambedkar movement and indicates a political and social awareness. Ambedkar adopted a different approach and philosophy for the emancipation of

Scheduled Castes. He wanted to liberate the dalits by building an egalitarian social order which he believed was not possible within the fold of Hinduism whose very structure was hierarchical which relegated the dalits to the bottom. Initially, he tried to seek emancipation of the dalits by bringing transformation within the structure of Hinduism through his efforts for opening the temples for the dalits and multi-caste dinners. However, Ambedkar came to realise soon that such an approach would not bring the desired result for the amelioration of the inhuman condition of the dalits. He asserted that the dalits should come forward and fight for their own cause. He gave them the *mantra* – educate, organise and agitate. He did not have faith in the charitable spirit of the caste Hindus towards the untouchables as it failed to bring any change in the oppressive social order. Ambedkar did not have any faith in Mahatmas and Saints whose main emphasis was not on the equality between man and man. Their philosophy, according to him, was mainly concerned with the relation between man and God.

Baba Saheb Dr. B. R. Ambedkar, himself a Dalit, made efforts to transform the hierarchical structures of Indian society for the restoration of equal rights and justice to the neglected lot by building up a critique from within the structure of Indian society. His was not a theoretical attempt but a practical approach to the problems of untouchability. He tried to seek the solution to this perennial problem of the Indian society not by making appeals to the conscience of the usurpers or bringing transformation in the outlook of the individual by begging but by seeking transformation in the socio-religious and politico-economic structures of the Indian society by continuous and relentless struggle against the exploitative system where he thought the roots of the untouchability lay. He thought that until and unless the authority of the *Dharamshastras* is shaken which provided divine sanction to the system of discrimination based on the case hierarchy, the eradication of untouchability could not be realised.

It was his subaltern perspective, a perspective from below which helped him to come to the conclusion that untouchability emanated neither from religious notions, nor from the much-popularised theory of Aryan conquest. He believed that it came into existence as a result of the struggle among the tribes at a stage when they were starting to settle down for a stable community living.

In the process, the broken tribesmen were employed by the settled tribes as guards against the marauding bands. These broken tribesmen employed as guards became untouchables. However, Ambedkar could not provide answer to the problem as to why only these broken tribesmen were confined to the one part of the village in the setting towards which the wind blew and the dirt of the village flowed. Ambedkar's tirade against untouchability was a tirade to make these people conscious of their rights, and to prepare them to agitate and win their rights.

Dr. B. R. Ambedkar and Indian Constitution: Dr. Ambedkar was one of the illustrious sons of the India who struggled throughout his life to restructure the Indian Society on humanitarian and egalitarian principles. He was not only a great national leader but also a distinguish scholar of international repute. He not only led various social movements for the upliftment of the depressed sections of the Indian Society but also contributed to the understanding of the Socio-Economic and Political problems of India through his scholarly works on caste, religion, culture, constitutional law and economic development. As a matter of fact he was an economist and his various scholarly works and speeches indicates his deep understanding of the problems faced by Indian Society.

From his very childhood, Dr. Ambedkar was forced to experience the evils of Untouchability. He being an untouchable by birth was forced to sit aside in classroom in the school. He was not allowed to touch his classmates. He could not mix and play with their fellows.

Owing to his academic brilliance, he won many government and private scholarships which helped him in attaining higher qualification from many Indian and foreign universities. He obtained a Ph.D. from Columbia in New York, D.Sc. from the London.

Having returned from London Dr. Ambedkar was given a high post in Baroda. When he reached Baroda no one came to welcome him. Worse still, even the servants in the office would not hand over the files to him. No one in the office would give him water to drink. He could not get a house to live in. On account of his low caste, he was refused a place on rent. He tried his best to get a room, but could find no shelter. He had to resign the post and returned to Bombay.

Dr. Ambedkar founded a Bahiskrit Hitkarini Sabha in 1924 for the upliftment of Dalits. He started Marathi fortnightly the "Mook Nayak" which means the leader of the dumb. Through this paper he awakened the dalits to fight for their rights. He arose a feeling of self respect and self confidence among the dalits and prepared for equal civil rights to the dalits.

Dr. Ambedkar launched marches to enter the Kala Ram Mandir in Nasik and drank water from the public tank at Mahad in Maharashtra. By this acts of the agitation, Dr. Ambedkar wanted to remove the mental dormancy of his people on an all India level. Dr. Ambedkar asked dalits to resist boldly all the acts of social tyranny" Goats! Lions are not sacrificed. Strengthen the organisation of the depressed classes all over the country as it is the only way for salvation."

In 1946, Dr. Ambedkar entered into Constituent Assembly. He was taken into the Drafting Committee and there after he was elected as the Chairman of the drafting Committee. He was also taken as the Minister for law.

Dr. Ambedkar knew the difficulties for his illiterate, gullible, resourceless and helpless people to carry on the massive struggles. The "orphans of the world" he resorted to the constitutional method to solve all the social, political and economic problems.

In the constitution, Dr. Ambedkar provided an inspiring preamble ensuring justice, social, economic and political, liberty, equality and fraternity. He provided comprehensive chapters of fundamental rights and opportunities without discrimination and for securing all freedom and for reducing economic inequalities. He also provided in the constitution safe guards for the protection of rights and interests of the dalits and minorities and equal rights to the women.

Dr. Ambedkar advocated his economic doctrine of "State socialism" in the draft constitution which he prepared and submitted to the constituent Assembly. He proposed state ownership of agriculture with a collectivised method of cultivation and a modified form of state socialism in the field of industry. Dr. Ambedkar also proposed that the above scheme of state socialism should not be left to the will of the legislature and it should be established by the law of constitution so that it will be beyond the reach of the parliamentary majority to suspend, amend or abrogate it. This guarantees state socialism while retaining parliamentary democracy. "It is only by this that one can achieve the triple objectives namely to establish socialism, retain parliamentary democracy and to avoid dictatorship".

Dr. Ambedkar also believed that this plan of state socialism was essential for increasing productivity without closing every avenue to private sector and also distributing wealth equitably. "The main purpose behind the clause is to put an obligation on the state to plan the economic life of people on lines which would lead to highest point of productivity without closing every avenue to the private enterprise and also provide for the equitable distribution of wealth." In other words his model of Economic Development was based on mixed economic system where state has to play active role while opening avenues for private sector.

Further Dr. Ambedkar also argued that if democracy was to live up to its principle of one man one value, it was essential to define both the economic structure as well as the political structure of the society by the law of constitution.

Thus it follows from the above analysis of Dr. Ambedkar's strategy of development that the state has to play very active and crucial role in accelerating the growth with justice through democratic methods. The foundation of democracy would be feeble and shaky if the contradictions between political democracy, enshrined in the constitution, and social and economic inequalities, existing in our society, are not resolved. And the shape and form of economic structure should be defined and prescribed by the law of the constitution without leaving it to the will of the legislature along with that of the structure of the political democracy so that social and economic democracy consistent with political democracy is established.

But due to the strong opposition of the constitution assembly, Dr. Ambedkar could not incorporate his scheme of state socialism under fundamental rights as a part of the constitution. Having failed to achieve his target by adopting such constitutional provisions for a free India, he resorted to the different mode to annihilate or at least weaken the caste system. He took a lengthy process of counter reservation.

The Constitution of India, the world's lengthiest written constitution was passed by the constituent assembly on November 26, 1949. It has been in effect since January 26, 1950, which is celebrated as Republic Day in India. Dr. Ambedkar said in the constituent assembly, "No matter how good a constitution may be, if the means to implement it are no good, then the constitution proves no better. If we want to establish democracy, then we must implement our social and economic ideal by means of peaceful and constitutional measures. Democracy's life is based on liberty, equality, and fraternity. There is a total lack of equality in India. We have equality in politics, but inequality reigns in the spheres of society and economics. Dr. Ambedkar's knowledge in the constitutional law was extensive varied, profound and encyclopedic. The role played by Dr. Ambedkar in the making of the Constitution received the appreciation of all the members of the constituent Assembly.

On 26th November, the Constituent assembly approved the draft Constitution Bill. Then the first President of India Dr. Rajendra Prasad praised the services rendered by Dr. Ambedkar in the making of the constitution and said:

> *"I have carefully watched the day to day activities from the Presidential seat. Therefore, I appreciate more than others with how much dedication and vitality this task has been carried out by the Drafting Committee and by its chairman Dr. Bhimrao Ambedkar in particular. We never did a better thing than having Dr. Ambedkar on the Drafting Committee and selecting him as its chairman."*

Columbia University as its Special convocation on 5th June 1952 conferred the LLD. degree (Honours Cause) on Dr. Ambedkar in recognition of his drafting the constitution of India. The citation read: "The degree is being conferred in recognition of the work done by him in connection with the drafting of India's constitution". The University hailed him as "one of India's leading citizen, a great social reformer and valiant upholder of human rights." The same university has already conferred the Ph.D. degree on Ambedkar thirty five years before.

Gautama Buddha fought to eradicate poverty, the Dukkha. He advocated the principle of no private property. He applied this principle of no private property within the limited campus of Sanghas. He advocated the doctrine of "Majjim Patipada" or the middle course. Men should neither accumulate nor enjoy too much or too little of wealth or privileges. That became the cause of "Dukkha" for the rest of the society. For the last and the latest social revolution Dr. Ambedkar himself adopted and advised his followers to adopt, Buddhism, the spirit of equality and justice. He took refuge in Lord Buddha on 14th October, 1956 at Nagpur along with 5 lakh of his followers. Baba Saheb, Ambedkar was planning to organise a mass conversion in very near future, but they were found dead on the night of 6th December 1956.

Important Events in His Life

- Birth, 14th April 1891
- Witness in South Barrow Commission, 1917
- Untouchables' conference, Nagpur, 1918
- Bahishkrit Hitkarini Sabha Formed 20th July 1924
- Nominated as MLC, Bombay Province, 1926
- Mahad Satyagraha, December 1927
- Witness in Simon Commission, May 1928
- Nashik Kala Ram temple Satyagraha, 2nd March 1930
- Representative at Round Table Conference, 1930-32
- British Communal Award, 20th August 1932
- Poona Pact, 20th September 1932
- Yewale District, Nasik Conference, 23rd October 1935
- Mahar Parishad, Bombay Province, 31st May 1936
- Independent Labout Party Formed, August 1936
- Elected MLA, Bombay Province, January 1937
- All India Scheduled Caste Federation Formed at Nagpur, April 1942

- Appointed as Labour Minister in the Viceroy's Executive Council, July 1942
- People's Education Society Formed, July 1945
- Elected to Constituent Assembly from Bengal, November 1946
- Law Minister in Independent India, 15th August 1947
- Appointed as Chairman, Drafting Committee of the Constitution of India, 29th August 1947
- Resigned from Union Cabinet, September 1951
- Elected to Rajya Sabha, March 1952
- Buddhist Society of India Formed, May 1955
- Embraced Buddhism, 14th October 1956
- Parinirvan (Death), 6th December 1956

Dr. Babasaheb Ambedkar and his Writings

- The National Dividend of India, 1916
- Small Holdings in India and Their Remedies, 1917
- Weekly 'Mook Nayak', Started 31st January 1920
- Provincial Decentralisation of Imperial Finance in British India, June 1921
- The Problem of a Rupee—Its Origin & Its Solution, March 1923
- The Evolution of Provincial Finance in British India, 1925
- Weekly 'Bahishkrit Bharat', Started 13th April 1927
- Weekly 'Janata', Started December 1930
- Annihilation of Caste, December 1935
- Federation Vs Freedom, January 1939
- Thoughts on Pakistan, December 1940
- Mr. Gandhi & the Emancipation of the Untouchables, December 1942
- Ranade, Gandhi & Jinnah, January 1943
- What Congress & Gandhi have done to the Untouchables, June 1945
- Who Were the Sudras?, October 1946
- States & Minorities, March 1947

- The Untouchable, October 1948
- Maharashtra as Linguistic Province, October 1948
- Thoughts on Linguistic States, December 1955
- Buddha & His Dhamma, Published 1957

22 Vows of Dr. Ambedkar

Dr. B.R. Ambedkar prescribed 22 vows to his followers during the historic religious conversion to Buddhism on 15 October 1956 at Deeksha Bhoomi, Nagpur in India. The conversion to Buddhism by 800,000 people was historic because it was the largest religious conversion, the world has ever witnessed. He prescribed these oaths so that there may be complete severance of bond with Hinduism. These 22 vows struck a blow at the roots of Hindu beliefs and practices. These vows could serve as a bulwark to protect Buddhism from confusion and contradictions. These vows could liberate converts from superstitions, wasteful and meaningless rituals, which have led to popularisation of masses and enrichment of upper castes of Hindus.

The famous 22 vows are:

- I shall have no faith in Brahma, Vishnu and Mahesh nor shall I worship them.
- I shall have no faith in Rama and Krishna who are believed to be incarnation of God nor shall I worship them.
- I shall have no faith in 'Gauri', Ganapati and other gods and goddesses of Hindus nor shall I worship them.
- I do not believe in the incarnation of God.
- I do not and shall not believe that Lord Buddha was the incarnation of Vishnu. I believe this to be sheer madness and false propaganda.
- I shall not perform 'Shraddha' nor shall I give 'pind-dan'.
- I shall not act in a manner violating the principles and teachings of the Buddha.
- I shall not allow any ceremonies to be performed by Brahmins.
- I shall believe in the equality of man.

- I shall endeavour to establish equality.
- I shall follow the 'noble eightfold path' of the Buddha.
- I shall follow the 'paramitas' prescribed by the Buddha.
- I shall have compassion and loving kindness for all living beings and protect them.
- I shall not steal.
- I shall not tell lies.
- I shall not commit carnal sins.
- I shall not take intoxicants like liquor, drugs etc.
- I shall endeavour to follow the noble eightfold path and practise compassion and loving kindness in every day life.
- I renounce Hinduism which is harmful for humanity and impedes the advancement and development of humanity because it is based on inequality, and adopt Buddhism as my religion.
- I firmly believe the Dhamma of the Buddha is the only true religion.
- I believe that I am having a rebirth.
- I solemnly declare and affirm that I shall hereafter lead my life according to the principles and teachings of the Buddha and his Dhamma.

Death

Since 1948, Ambedkar had been suffering from diabetes. He was bedridden from June to October in 1954 owing to clinical depression and failing eyesight. He had been increasingly embittered by political issues, which took a toll on his health. His health worsened as he furiously worked through 1955. Just three days after completing his final manuscript 'Buddha And His Dhamma', it is said that Ambedkar died in his sleep on December 6, 1956 at his home in Delhi.

Some of his supporters doubt his natural death and think death may not have been natural and put forward different theories. Try to draw parallel between his and Gandhi's death and blame Caste Brahmins. A Buddhist-style cremation was organised for him at Chowpatty beach on December 7, attended by hundreds of thousands of supporters, activists and admirers.

Ambedkar was survived by his second wife Savita Ambedkar, born as a Caste Brahmin and converted to Buddhism with him. His wife's name before marriage was Sharda Kabir. Savita Ambedkar died as Buddhist in 2002. Ambedkar's grandson, Prakash Yaswant Ambedkar leads the Bharipa Bahujan Mahasangha and has served in both houses of the Indian Parliament.

A number of unfinished typescripts and handwritten drafts were found among Ambedkar's notes and papers and gradually made available. Among these were *Waiting for a Visa*, which probably dates from 1935-36 and is an autobiographical work, and the *Untouchables, or the Children of India's Ghetto*, which refers to the census of 1951.

A memorial for Ambedkar was established in his Delhi house at 26 Alipur Road. His birthdate is celebrated as a public holiday known as Ambedkar Jayanti. He was posthumously awarded India's highest civilian honour, the Bharat Ratna in 1990. He is the namesake of many public institutions, such as the Dr. Babasaheb Ambedkar Open University in Ahmedabad, Gujarat.

A large official portrait of Ambedkar is on display in the Parliament building. Jabbar Patel directed the Hindi-language movie "Dr. Babasaheb Ambedkar" about the life of Ambedkar, released in 2000, starring South Indian actor Mammootty as Ambedkar. Sponsored by India's National Film Development Corporation and Ministry of Social Justice, the film was released after a long and controversial gestation period. Dr. Ambedkar was the main architect of the Indian Constitution. He was born in a very poor low caste family of Madhya Pradesh. In U.S.A., he did his M.A. in 1915 and Ph.D. in 1916. From 1918 to 1920, he worked as a Professor of Law. Dr. Ambedkar set up his legal practice at the Mumbai High Court.

A number of unfinished typescripts and handwritten drafts were found among Ambedkar's notes and papers and gradually made available. Among these were Waiting for a Visa, which probably dates from 1935-36 and is an autobiographical work, and the Untouchables, or the Children of India's Ghetto, which refers to the census of 1951. A memorial for Ambedkar was established in his Delhi house at 26 Alipur Road. His birthdate is celebrated as a public holiday known as Ambedkar Jayanti. He was posthumously awarded India's highest civilian honour, the Bharat Ratna in 1990. He is the namesake of many public institutions, such as the Dr. Babasaheb Ambedkar Open University in Ahmedabad, Gujarat. A large official portrait of Ambedkar is on display in the Parliament building. Jabbar Patel directed the Hindi-language movie "Dr. Baba Saheb Ambedkar" about the life of Ambedkar, released in 2000, starring South Indian actor Mammootty as Ambedkar. Sponsored by India's National Film Development Corporation and Ministry of Social Justice, the film was released after a long and controversial gestation period.

Efforts to End Injustice

There was no cast system during the Vedic age. There was no 'untouchability'. When and how did this system creep into the Hindu society? We do not know for certain.

Did no one try to wipe out this injustice? Buddha admitted may 'untouchables' to his religion. Ramanujacharya, Basaveshwara, Chakradhara, Ramananda, Kabir, Chaitanya, Ekanath, Tukaram, Raja Rammohan Roy and other great men preached that no one is high and no one is low among God's children. Mahatma Phooley and his wife dedicated their lives to the education of the 'untouchables.' Sayyaji Rao Gaikwad, the Maharaja of Baroda, established a school for the 'untouchables' as early as in 1883. In this way many thoughtful leaders of the Hindu Society have been trying for hundreds of years to wipe out 'Untouchability'. Both before and after India became free, many great man have sacrificed their lives for the truth and the principles they believed in. Ambedkar was one of them. Ideas of high and low had crept into the Hindu Society; Ambedkar suffered because of this; he also fought hard against such differences; later he became the first Las Minister in free India. The credit for making a law and creating the necessary atmosphere to wipe out 'Untouchability' goes to Ambedkar.

Should Not Hindu, Who Seek Justice, Give Justice?

The 'untouchables' are Hindus. Therefore, the doors of temples should be open to them. If the Hindus can touch the Christians and the Muslims, why should they not touch the people who are themselves Hindus and who worship the Hindu Gods? This was Ambedkar's argument. He gave a call that people who practise and support 'untouchability' should be punished. Some people argued that the 'untouchables' were not yet fit for equality. The Hindus say that they want independence and democracy. How can a people who have temple upon all the libertise of a backward group aspire to democracy? Ambedkar argued like this and thundered that these people had no right to speak of justice and democracy.

In 1927 there was a big conference. It resolved that there should be no cast differences in the Hindu Dharma and that people of all castes should be allowed to work as priests in temples. The Chowdar Tank dispute went to the court. The court decided that tanks are public property. The 'untouchables' who have been subjected to humiliation for hundreds of years should find justice. For this purpose Ambedkar indicated a few clear steps. No section of the Hindus should be kept out of temples.

There should be more representatives of the 'untouchables' in the legislatures. These representatives should not be nominated by the government. They should be elected by the people. The government should employ the 'untouchalbes' in larger numbers the army and the police department.

A Fearless, Firm Mind: Those who suffer in the Hindu Society should get justice.

This was Ambedkar's rocklike decision. He was prepared to oppose anybody to reach his goal. The British Government invited several Indian leaders to discuss the problems of India. The conferences were held in London; they were called the 'Round Table Conference'. Gandhiji also took part in them. At the Round Table Conference Ambedkar spoke angrily against the government. He said that the backward sections did not enjoy equality with other sections, even under the British Government; the British had just followed the ways of the other Hindus.

This was a time when Gandhiji was very popular in India. Millions of people followed his footsteps with devotion. Ambedkar openly opposed Gandhiji's views on how justice should be secured for the 'untouchables'. He supported the views which seemed right to him. Ambedkar secured for the Harijans (the 'untouchables') 'separate electorates' at the Second Round Table Conference in 1931. As a result, the Harijans could elect their representatives separately.

After Resignation as a Minister

In 1952, he was defeated by a Congress candidate in the election for the Lok Sabha. The entire country was shocked by his defeat.

Later he was elected to the Rajya Sabha. Whenever he felt that the government had not done justice to the Harijans he criticised it sharply. In 1953 the government brought a bill before the parliament. According to this bill those who practised 'untouchablity' would be punished; imprisonment, imposition of fines, dismissal from employment and withdrawal of licence to follow a profession – these were the forms of punishment.

To the Path of the Buddha: Soon after the framing of the Constitution, Ambedkar's mind turned towards Buddha. His mind was thirsting for peace and justice. He attended the Buddhist Conference in Ceylon (Sri Lanka) in 1950.

The bitterness of his mind was ever on the increase. In spite of it, he was not willing to embrace the Christian or the Muslim faith.

Finally, Ambedkar decided to become a Buddhist. This was a great decision in his life, a decision taken after deep thought.

A Neglected Message from Dr. Ambedkar to OBCs

An article was published recently in Marathi local magazine by Suhas Sonwane based on daily Loksatta. The following is a gist of it, translated from Marathi.

Mr. Baba Saheb Gawande, the founder president of an Organization of Marathas from Bombay called "Maratha Mandir" was a close friend of Dr. Ambedkar. Mr. Gawande asked Dr. Ambedkar, who was then a Law Minister in Nehru Cabinet in 1947, for a message for the Maratha people to be published in the Souvenir of "Maratha Mandir". Ambedkar declined saying that he had no relation with the Organization or the Marathas, but on persistent insistence, a message was given and published in the souvenir on 23rd March 1947. Unfortunately that special issue is not available in the office of the Organization today, but was made available by Shri Vijay Survade recently and was undocumented till now.

Dr. Ambedkar said:

"This principle will apply not only to Marathas but all Backward Castes. If they do not wish to be under the thumb of others they should concentrate on two things, one is politics and the other is education."

"One thing I like to impress on you is that the community can live in peace only when it has enough moral but indirect pressure over the rulers. Even if a community is numerically weak, it can keep its pressure over the rulers and create its dominance as is seen by the example of status of present day Brahmins in India. It is essential that such a pressure is maintained, as without it, the aims and policies of the state can not have proper direction, on which depends the development and progress of the state."

"At the same time, it must not be forgotten that education is also important. Not only elementary education but higher education is most essential to keep ahead in competition of communities in their progress."

"Higher education, in my opinion, means that education, which can enable you to occupy the strategically important places in State administration. Brahmins had to face a lot of opposition and obstacles, but they are overcoming these and progressing ahead."

"I can not forget, rather I am sad, that many people do not realize that the Caste system is existing in India for centuries because of inequality and a wide gulf of difference in education, and they have forgotten that it is likely to continue for some centuries to come. This gulf between the education of Brahmins and non-Brahmins will not end just by primary and secondary education. The difference in status between these can only be reduced by higher education. Some non-Brahmins must get highly educated and occupy the strategically important

places, which has remained the monopoly of Brahmins since long. I think this is the duty of the State. If the Govt. can not do it, institutions like "Maratha Mandir" must undertake this task."

"I must emphasize one point here that middle class tries to compare itself with the highly educated and well placed and well to do community, whereas lower class all over the world has same fault. The middle class is not as liberal as upper one, and has no ideology as lower one, which makes it enemy of both the classes. The middle class Marathas of Maharashtra also have this fault. They have only two ways out, either to join hands with upper classes and prevent the lower classes from progress, and the other is to join hands with lower classes and both together destroy the upper class power coming against the progress of both.

There was a time, they used to be with lower classes, now they seem to be with the upper class. It is for them to decide which way to go. The future of not only Indian masses but also their own future depends upon what decision the Maratha leaders take. As a matter of fact it all should be left to the skill and wisdom of the leaders of Marathas. But there seems to be a lack of such wise leadership among the Marathas."

What he said about Marathas, equally applies to all OBCs, and still holds true after half a century. Dr. Ambedkar wrote much to educate the OBCs. It is only now that OBCs are awakening gradually. It must not be forgotten that the future of this country depends on them.

3

Ambedkar on Tyranny of Hindus

Rules for Balais

Mode of Life Laid Down: "Last May (1927) High Caste Hindus, *viz.* Kalotas, Rajputs and Brahmins, including the patels and putwaris of villages Kanaria, Bicholee Hafsi, Mardana and of about 15 other villages in the Indore District, informed the Balais of their respective village that if they wished to live among them, they must conform to the following rules: (1) Balais must not wear gold lace bordered pugrees; (2) they must not wear dhoties with coloured or fancy borders; (3) they must convey intimation of the death of any Hindu to relatives of the deceased—no matter how far away these relatives might be living; (4) in all Hindu marriages, the Balais must play music before the processions, and during the marriage; (5) the Balai women must not wear gold or silver ornaments; they must not wear fancy gowns, or jackets; (6) Balai women must attend all cases of confinement of Hindu women; (7) the Balais must render services without demanding remuneration, and must accept whatever a Hindu is pleased to give; (8) if the Balais do not agree to abide by these terms, they must clear out of the villages.

Balais refuse Compliance

The Balais refused to comply; and the Hindu element proceeded against them. Balais were not allowed to get water from the village wells, they were not allowed to let their cattle graze. Balais were prohibited from passing through land owned by a Hindu; so that if the field of a Balai was surrounded by fields owned by Hindus, the Balai could have no access to his own field. The Hindus also led their cattle to graze down the fields of Balais. The Balais submitted petitions to the Darbar against these persecutions; but as they could get no timely relief, and the operation continued, hundreds of Balais, with their wives and children, were obliged to abandon their homes in which their ancestors lived for generations and migrate to adjoining States, *viz.*, to villages in Dhar, Dewas, Bhopal, Gwalior and other States.

Compulsory Agreement

Only a few days ago the Hindus of Reoti village, barely seven miles to north of Indore City, ordered the Balais to sign a stamped agreement in accordance with the rules framed against the Balais by the Hindus of other villages. The Balais refused to comply. It is alleged that some of them were beaten by the Hindus; and one Balai was fastened to a post, and was told that he would be let go on agreeing to sign the agreement. He signed the agreement and was released. Some Balais from this village ran up to the Prime Minister the next day, *i.e.* on the 20th December, and made a complaint about the ill treatment they received from the Hindu villagers of Reoti. They were sent to the Subha of the district. This officer, with the help of the police, made inquiries at the village, and recommended that action be taken against the Hindus under section 342 and 147 and against the Balais under section 147, Indian Penal Code.

Balais Leave Villages

Caste Tyranny

Ignorance of Law a Handicap: "There has been no improvement in the treatment of the Balais by the Hindu residents of certain villages. Balais, it has already been reported, have been ill treated by the higher caste Hindus. From the Dopalpur Pargana alone, Indore District, a large number of Balais have had to leave their homes and find shelter in adjoining States. The villages from which Balais have been forced to clear out are Badoli, Ahirkharal, Piploda, Morkhers, Pamalpur, Karoda, Chatwada, Newri, Pan, Sanauda, Ajnoti, Khatedi and Sanavada. Pamalpur village has been altogether deserted and not a Balai man, woman or child is to be found there. Nanda Balai a resident of one of the above villages, it is alleged, was severely beaten by the Hindus of the village. In one village, the report goes, the Hindus burnt down all the dwellings of the Balais but the offenders have not yet been traced.

"Balais are ignorant village folk, who are ignorant of legal procedure and think that if a petition is sent to the Sirkar all that is required will be done for them. They have not the knowledge; or the means and practices, to pursue a complaint to its end; and, as they, it is said in some cases, failed to attend or produce witnesses in support of their allegations, the magistrate had no alternative but to dismiss their complaint."

Looked at from the point of view of Dharma and Adharma, can it be doubted that underneath the lawlessness and ruthlessness of the Hindus in suppressing the revolt of the untouchables, they are actuated by what they think a noble purpose of preventing an outrage upon their Dharma?

It may well be asked how much of this Dharma of Manu now remains? It must be admitted that as law in the sense of rules which a Court of Judicature is bound to observe in deciding disputes, the Dharma of Manu has ceased to have any operative force-except in matters such as marriage succession etc. matters which affect only the individual. As Law governing social conduct and civic rights it is inoperative. But if it has gone out as law, it remains as custom.

Custom is no small a thing as compared to Law. It is true that law is enforced by the state through its police power; custom, unless it is valid it is not. But in practice this difference is of no consequence. Custom is enforced by people far more effectively than law is by the state. This is because the compelling force of an organized people is far greater than the compelling force of the state.

Not only has there been no detriment to its enforceability on account of its having ceased to be law in the technical sense but there are circumstances which are sufficient to prevent any loss of efficacy to this Dharma of Manu.

Of these circumstances the first is the force of custom. There exists in every social group certain (habits) not only to acting, but of feeling and believing, of valuing, of approving and disapproving which embody the mental habitudes of the group. Every new comer whether he comes in the group by birth or adoption is introduced into this social medium. In every group there goes on the process of persistently forcing these mental habitudes of the group upon the attention of each new member of the group. Thereby the group carries on the socialization of the individual of the shaping of the mental and practical habits of the new comer. Being dependent upon the group he can no more repudiate the mental habitudes of the group than he can the condition and regulation of his physical environment. Indeed, so dependent the individual is on the group that he readily falls in line and allows the current ways of esteeming and behaving prevailing in the community, to become a standing habit of his own mind. This socializing process of the individual by the group has been graphically described by Grote. He says:

> *"This aggregate of beliefs and predispositions to believe, ethical, Religious, Aesthetical, and Social respecting what is true, or false, probable or improbable, just or unjust, holy or unholy, honourable or base, respectable or contemptible, pure or impure, beautiful or ugly, decent or indecent, obligatory to do, or obligatory to avoid, respecting the status and relations of each individual in the society, respecting even the admissible fashions of amusement and recreation—this is an established fact and condition of things, the real origin of which for the most part unknown, but which each new member of the group is born to and finds subsisting..... It becomes a part of each person's nature, a standing habit of mind, or fixed set of mental tendencies, according to which particular experience is interpreted and particular persons appreciated..... The community hate, despise or deride any individual member who proclaims his dissent from their social creed..... Their hatred manifests itself in different ways..... At the very best by exclusion from that amount of forbearance, good will and estimation*

without which the life of an individual becomes insupportable."

But what is it that helps to bring about this result? Grote has himself answered this question. His answer is that, this is due to— "Nomos (Law and Custom), King of all "(which Herodotus cites from Pindar) exercises plenary power, spiritual and temporal, over individual minds, moulding the emotions as well as the intellect, according to the local type.... and reigning under the appearance of habitual, self suggested tendencies.

What all this comes to is that, when in any community, the ways of acting, feeling, believing, or valuing or of approving and disapproving have become crystallised into customs and traditions, they do not need any sanction of law for their enforcement. The amplitude of plenary powers which the group can always generate by mass action is always ready to see that they are not broken.

The same thing applies to the Dharma laid down by Manu. This Dharma of Manu, by reason of the governing force which it has had for centuries, has become an integral and vital part of the customs and traditions of the Hindus. It has become ingrained and has given colour to their life blood. As law it controlled the actions of the Hindus. Though now a custom, it does not do less. It moulds the character and determines the outlook of generation after generation.

The second thing which prevents the Dharma of Manu from fading away is that the law does not prevent its propagation. This is a circumstance which does not seem to be present to the minds of many people. It is said that one of the blessings of the British Rule is that Manusmriti has ceased to be the law of the land. That the Courts are not required to enforce the provisions contained in Manusmriti as rules of law is undoubtedly a great blessing—which might not be sufficiently appreciated except by those who were crushed beneath the weight of this "infamous" thing. It is as great a blessing to the untouchables as the Reformation was to the peoples of Europe. At the same time it must be remembered that the Reformation would not have been a permanent gain if it had been followed by what is called the Protestant Revolution.

The essential features of the Protestant Revolution as I understand them are:

(1) That the state is supreme and the Church is subordinate to the state.

(2) The doctrine to be preached must be approved by the state.

(3) The clergy shall be servants of the state and shall be liable to punishment not only for offences against the general law of the land but also liable for offences involving moral turpitude and for preaching doctrines not approved of by the state. I am personally a believer in the "Established Church". It is a system which gives safety and security against wrong and pernicious doctrines preached by any body and every body as doctrines of religion. I know there are people who are opposed to the system of an "Established Church". But whether the system of an "Established Church" is good or bad, the fact remains that there is no legal prohibition against the propagation of the Dharma laid down by Manu. The courts do not recognize it as law. But the law does not treat it as contrary to law. Indeed every village every day. When Pandits are preaching it to parents and parents preach it to their children, how can Manusmriti fade away? Its lessons are reinforced every day and no body is allowed to forget that untouchability is a part of their Dharma.

This daily propagation of the Dharma of Manu has infected the minds of all men and women young and old. Nay, it has even infected the minds of the judges. There is a case reported from Calcutta. A certain Dome (untouchable) by name Nobin Dome was prosecuted for theft of a goat. He was found to be not guilty. He filed a complaint for defamation against the complaint. The magistrate dismissed the complaint on the ground that as he was low caste man he had no reputation. The High Court had to intervene and direct the Magistrate that he was wrong in his view and that under the Penal Code all persons were equal. But the question remains, how did the Magistrate get the idea that an untouchable had no reputation? Surely from the teaching of the Manusmriti.

The Dharma of Manu had never been a mere past. It is as present as though it were enacted today. It bids fair to continue to have its sway in the future. The only question is whether its sway will be for a time or forever.

Under the Providence of Mr. Gandhi

(1) His work through the Congress

1. A Strange Welcome.
2. The Great Repudiation.
3. A Charge Sheet.
4. The Basis of the Charge Sheet.
5. The Tragedy of Gandhi.
6. His Legacy to India and the Untouchables.

On the 28th December, 1931, Mr. Gandhi returned to India from London where he had gone as a delegate to attend the second Session of the Indian Round Table Conference. At the Round Table Conference, Mr. Gandhi had been an utter, ignominious failure both as a personality and as a politician. I know that my opinion will not be accepted by the Hindus. But the unfortunate part is that my opinion in this respect coincides with the opinion of Mr. Gandhi's best friend. I will cite the opinions of two. This is what Mr. Ewer, who was closely associated with Mr. Gandhi during the Round Table Conference, wrote about the role Mr. Gandhi played at the Round Table Conference in London.

> *"Gandhi in the St. James Palace has not fulfilled the unwise expectation of those who saw him bestriding the Conference like a colossus........... He was out of his elements."*
>
> *"His first speech, with its sentimental appeal, its over-stressing of humility, its reiteration of single-minded concern for the dumb suffering millions, was a failure. No one questioned its sincerity. But somehow it rang false. It was the right thing, perhaps, but it was in the wrong place. Nor were his later interventions on the whole more successful. A rather querulous complaint that the British Government had not produced a plan for the new Indian Constitution shocked some of*

Gandhi's colleagues, who had hardly expected to see the representative of the National Congress appealing to British Ministers for guidance and initiative. The protest against the pegging of the rupee to the pound was astonishingly ineffective. The contributions to the discussions on franchise and kindred matters were of little importance. Behind the scenes he was active enough in the Hindu-Moslem negotiations, but here, too, results were intangible. Not for a moment did Gandhi take the lead or materially influence the course of committee work. He sat there, sometimes speaking, sometimes silent, while the work went on, much as it would have gone on without him."

This is what Bolton has to say about Mr. Gandhi's achievement at the Round Table Conference.

How did Mr. Gandhi fare as a statesman and a politician? At the close of the first session of the Round Table Conference there were three questions which had not been settled. The question of minorities, the question of the Federal structure and the question of the status of India in the Empire, were the three outstanding problems, which were the subject matter of controversy. Their solution demanded great statesmanship. Many said that these questions were not settled because the wisdom and authority of the Congress was not represented at the Round Table Conference. At the second session, Mr. Gandhi came and made good the deficiency. Did Mr. Gandhi settle any of these unsettled problems? I think it is not unfair to say that Mr. Gandhi created fresh disunity in the Conference.

He began the childish game of ridiculing every Indian delegate. He questioned their honesty, he questioned their representative character. He taunted the liberals as armchair politicians and as leaders without any followers. To the Muslims he said that he represented the Muslim masses better than they did. He claimed that the Depressed Class delegates did not represent the Depressed Classes and that he did. This was the refrain which he repeated *ad nauseum at* the end of every speech.

The non-Congress delegates deserve the thanks of all honest people for their having tolerated this nonsense and arrogance of Mr. Gandhi and collaborated with him to save him and to save the country from his mistake. Apart from this discourtesy to fellow-delegates, did Mr. Gandhi stand up for the cause he came to champion? He did not. His conduct of affairs was ignominious. Instead of standing up and fighting he began to yield on issues on which he ought never to have ceased fire. He yielded to the Princes and agreed that their representatives in the Federal legislature should be nominated by them and not elected, as demanded by their subjects. He yielded to the conservatives and consented to be content with provincial autonomy and not to insist upon central responsibility for which many lakhs of Indians went to gaol. The only people to whom he would not yield were the minorities — the only party to whom he could have yielded with honour to himself and advantage to the country.

Nothing has helped so much to shatter the prestige of Mr. Gandhi as going to the Round Table Conference. The spectacle of Mr. Gandhi at the Round Table Conference must have been painful to many of his friends. He was not fitted to play the role he undertook to play. No country has ever sent a delegate to take part in the framing of the constitution who was so completely unequipped in training and in study. Gandhi went to the Round Table Conference with a song of the saint Narsi Mehta on his tongue. It would have been better for him and better for his country if he had taken in his arm pit a volume on comparative constitutional law. Devoid of any knowledge of the subject he was called upon to deal with, he was quite powerless to destroy the proposals put forth by the British or to meet them with his alternatives. No wonder Mr. Gandhi, taken out of the circle of his devotees and placed among politicians, was at sea. At every turn he bungled and finding that he could not even muddle through, he gave up the game and returned to India.

How was Mr. Gandhi received when he landed on the Indian soil? It may sound strange to outsiders and to those who are not the devotees of Mr. Gandhi but it is a fact that when the *S. S. Pilsner* of the Lloyd Triestino entered the harbour of Bombay at 8 a.m. in the morning of the 28th December 1931 there came to receive him an enthusiastic crowd of men, women and children who had assembled at the Pier in tens of thousands to greet him, to welcome him back and to have his *Durshan*. The following extracts from the *Times of India* and the *Evening News* of Bombay will serve to give a vivid idea of the grandeur of this reception.

The Pilsner was escorted into the harbour by Desh Sevikas (women volunteers of the Congress) in saffron coloured sarees who went out in launches some distance from the pier. The Congress Committee had asked the Bombay Flying Club to fly an Aeroplane or two over the Pilsner and drop garlands as she came along side the pier, but the Flying Club, sanely preferring to keep out of politics, refused to grant the Congress demand.

The spacious Central Hall at Ballard Pier was decorated with festoons and Congress flags and a large dais was put up at the centre with chairs placed on all sides for representatives of various organisations, local and upcountry, who were given passes for admission. Both the approaches to the reception hall from the wharf and from the city were lined by Desh Sevikas waving national flags and the duty of guarding the dais and of regulating and directing the assembly inside the hall was also entrusted to the women volunteers.

Mr. Gandhi reached the dais escorted by the Congress leaders and received an ovation. Hardly had he stepped on the dais when he began to be flooded with telegraph messages (presumably of welcome) which arrived one after another.

Standing on the dais he was garlanded in turn by representatives of the public bodies who had assembled and whose names were called out from a long printed list of which copies were previously distributed.

The proceedings inside the reception hall terminated with the garlanding.

A procession was then formed in four, in place of the carriage which was intended to be the conveyance for Mr. Gandhi. He was seated in a gaily decorated motor car, with Mr. Vallabhbhai Patel to his left and Mr. Vitthalbhai Patel to his right and Mr. K. F. Nariman, President of the Bombay Provincial Congress Committee, on the front seat.

Preceded by a pilot car and followed by others containing the Congress Working Committee members, the procession passed through the Ballard Pier Road, Hornby Road and

Kalbadevi which had been decorated by the citizens at the instance of the Congress Committee and lined on either side by cheering crowds, five to ten deep till the party reached "*Mani Bhuvan,* Gamdevi". At no stage of this welcome did Mr. Gandhi open his lips to acknowledge it. This man of vows was under his Sunday vow of silence which had not run out till then and nor did he think that etiquette, good manners or respect for those who had assembled required that he should terminate his vow earlier.

The official historian of the Congress describes' this reception given to Mr. Gandhi in the following terms:

> *"There were gathered in Bombay representatives of all parts and Provinces in India to accord a fitting welcome to the Tribune of the people. Gandhi greeted the friends that went on board the steamer to welcome him, patting many, thumping a few and pulling the venerable Abbas Tyabji by his beard. There was a formal welcome in one of the Halls of Customs House and then a procession in the streets of Bombay which kings might envy in their own country".*

On reading this account one is reminded of the Irish Sein Fein Delegates who in 1921, just 10 years before, had gone to London at the invitation of Mr. Lloyd George for the settlement of the Irish Home Rule question. As is well known the Irish Delegates secured from the British Cabinet a treaty which was signed on the 8th December 1921. The Treaty was subsequently submitted for approval to the Dail, the Parliament of the Sein Fein Party which met from 14th December 1921 to 7th January 1922. On the 7th January a division was taken. There were 64 votes for ratifying the treaty and 57 against. And what was the reception given to the Irish delegates who secured this treaty? Arthur Griffith— who was the head of the Irish Delegation and Michael Collins who was his most prominent colleague, were both of them shot by the anti-treaty Sein Feiners, the former on the 12th and the latter on the 22nd August 1922. The reason for sending them to such cruel death was that the treaty which they signed did not secure the inclusion of Ulster and a republic for Ireland. It is true the treaty did not grant this. But if it is remembered that negotiations were opened on the express understanding on the part of both sides that these two questions were outside the scope of negotiations it will be granted that if the treaty did not include these it was no fault of the Irish Delegates. The fury and ferocity of the anti treaty Sein Feiners against the Irish Delegates had no moral foundation and the fate that befell Arthur Griffith and Michael Collins can by no stretch of imagination be said to be one which they deserved.

Be that as it may, this welcome to Mr. Gandhi will be regarded as a very strange event. Both went to win Swaraj, Griffith and Collins for Ireland, Gandhi for India, Griffith and Collins succeeded, almost triumphed; Gandhi failed and returned with nothing but defeat and humiliation. Yet Collins and Griffith were shot and Gandhi was given a reception which kings could have envied!! What a glaring and a cruel contrast between the fate that awaited Collins and Griffith and the reception arranged for Gandhi? Are the Indian Patriots different from the Irish Patriots? Did the masses render this welcome out of blind devotion or were they kept in darkness of the failure of Mr. Gandhi by a mercenary Press? This is more than I can answer.

While this great welcome was being accorded to Mr. Gandhi the Untouchables of Bombay had come to the Pier to repudiate.

Mr. Gandhi. Referring to this demonstration, the newspaper reports said:

> *"Just outside the gate of Ballard Pier, the scene was most exciting. On one side were drawn up Depressed Class volunteers in uniform, weaving black flags to the accompaniment of derisive shouts against Mr. Gandhi and laudatory cries in praise of their leader, while on the other side Congress followers kept up a din of counter shouts."*

This Untouchable demonstration included men and women. The demonstrators numbered thousands, all waving Black Flags as a mark of repudiation of Mr. Gandhi. They were a determined crowd and, despite intimidation by the superior forces of the Congress assembled there to welcome Mr. Gandhi, were bent on showing that they repudiated Mr. Gandhi. This led to a clash and blood was split. There were forty casualties on each side. For the first time Mr. Gandhi was made aware that there could be black flags even against him. This must have come to him as a shock. When he was asked about it later in the day, he said he was not angry, the Untouchables being the flesh of his flesh and bone of his bone. This is of course the Mahatmaic way of concealing the truth. One would not mind this convenient and conventional lie if there were behind it a realization that the crowd could not always be trusted to be loyal to its hero. Congressmen in India sadly lack the realism of a man like Cromwell. It is related that when Cromwell returned after a great battle, an enormous crowd came out to greet him. A friend sought to impress upon him the immensity of the crowd. But Cromwell dismissed the subject with the leconic remark, "Oh yes, I know many more will come to see me hanged."[1] No Congress leader feels the realism of Cromwell. Either he believes that the day will never come when he will be hanged or he believes that the Indian crowd will never become a thinking crowd. That part of the Indian crowd does think was shown by the representatives of the Untouchables who assembled on the 28th to greet Mr. Gandhi with black flags.

Why did the Untouchables repudiate Mr. Gandhi? The answer to this question will be found in a statement issued by the organizers of these demonstrations which was printed and circulated on that day. The following are extracts from it. "Our Charge sheet against Gandhiji and Congress", "Enough of patronising attitude and lip sympathy. We ask for justice and fair play."

1. In spite of the fact that the removal of untouchability has been included in the constructive programme of the Congress, practically nothing has so far been done by that body to achieve that object, and in our fights against untouchability at Mahad and Nasik most of the local Congress leaders have been our bitter opponents.
2. The attitude of Gandhiji at the Round Table Conference in London with regard to the demands of the Depressed Classes as put forward by their accredited and trusted leader Dr. Ambedkar, was most unreasonable, obstinate and inexplicable.
3. Gandhiji was prepared to concede on behalf of the Congress the special claims of the Mohamedans and the Sikhs including their demand for separate representation on "historic grounds", but he was not

willing even to concede reserved seats in general electorates to the Depressed Classes, although he knew, or should have known, what sort of treatment they would get, should they be thrown upon at the mercy on caste Hindus.

4. Gandhiji has said in opposing the claims of the Depressed Classes for separate representation that he does not want the Hindu Community to be subjected to vivisection or dissection. But the Congress is now dissecting the community of Untouchables by playing one section against another. Gandhiji and the Congress are not playing the fair game. Open enemies are far better than treacherous friends.
5. Attempts are being made to show that Gandhiji and the Congress alone represent the Depressed Classes by presenting addresses through a handful of hirelings and dupes. Is it not our duty to demonstrate the fact by coming out in thousands and proclaiming the truth? This is our charge sheet against Gandhiji and the Congress.

Let those who are not blind hero worshippers and blind partisans judge and give their verdict.

Is this charge sheet true? Mr. Gandhi is known to the world not merely as the Political leader of India, but also as the Champion of the Untouchables. It is perhaps true that the outside world takes more interest in Mr. Gandhi because he is the champion of the Untouchables than because he is a political leader. For instance the Manchester Guardian very recently devoted an editorial to the work of Mr. Gandhi for the Untouchables. In the face of this, the charge appears to be quite unfounded. For, has not Mr. Gandhi made the Congress pledge itself to remove untouchability? The Congress before it came into the hands of Mr. Gandhi had refused to allow any social problem to be placed before it for consideration. A clear cut distinction was made between political and social question, and scrupulous attempt was made to confine the deliberations and activities of the Congress to purely political questions. The old Congress refused to take notice of the Untouchables. It was with great difficulty that the Congress in 1917 for the first time allowed the question of the Untouchables to be placed before it and condescended to pass the following resolution:

"The Congress urges upon the people of India the necessity, justice and righteousness of removing all disabilities imposed by custom upon the Depressed Classes, the disabilities being of a most vexatious and oppresssive character, subjecting those classes to considerable hardship and inconvenience."

The Congress fell onto the hands of Mr. Gandhi in 1920 and the Congress at its ordinary session held at Nagpur passed the following resolution:

Intercommunal Unity

"Finally, in order that the Khilafat and the Punjab wrongs may be redressed and Swarajya established within one year, this Congress urges upon all public bodies, whether affiliated to the Congress or otherwise, to devote their exclusive attention to the promotion of non-violence and non-cooperation with the Government and, inasmuch as the movement of non-cooperation can only succeed by complete cooperation amongst the people themselves, this Congress calls upon public associations to advance Hindu-Muslim unity and the Hindu delegates of this Congress call upon the leading Hindus to settle all disputes between Brahmins and Non-Brahmins, wherever they may be existing, and to make a special effort to rid Hinduism of the reproach of untouchability, and respectfully urges the religious heads to help the growing desire to reform Hinduism in the matter of its treatment of the suppressed classes."

Again did not Mr. Gandhi make the removal of untouchability a condition precedent for achieving Swaraj? In the *Young India* of December 29, 1920, Mr. Gandhi wrote:

"Non-cooperation against the Government means cooperation among the governed, and if Hindus do not remove the sin of untouchability, there will be no Swaraj in one year or one hundred years...."

Writing again on the conditions of Swaraj in the issue of *Young India* for February 23, 1921, he said:

"Swaraj is easy of attainment before October next if certain simple conditions can be fulfilled. I ventured to mention one year in September last because I knew that the conditions were incredibly simple and I felt that the atmosphere in the country was responsive. The past five months experience has confirmed me in the opinion. I am convinced that the country has never been so ready for establishing Swaraj as now."

"But what is necessary for us as accurately as possible to know the conditions. One supreme indispensible condition is the continuance of non-violence."

"The next condition is...... establishing a Congress Agency in every village."

"There are certain things that are applicable to all. The potent thing is Swadeshi. Every home must have the spinning wheel and every village can organize.... and become self supporting."

"Every man and woman can give some money—be it even a pice—to the Tilak Swaraja Fund. And we need have no anxiety about financing the movement."

"We can do nothing without Hindu-Moslem unity and without killing the snake of untouchability."

"Have we honest, earnest, industrious, patriotic workers for this very simple programme? If we have, Swaraj will be established in India before next October."

What more did the Untouchables want? Here is Mr. Gandhi who had held himself out as the friend of the Untouchables. He prides himself on being their servant. He claims and fought for being accepted as their representative. Why should the Untouchables show such a lack of confidence in Mr. Gandhi? On the basis of words, the charge perhaps appears unfounded. But does it appear equally unfounded if we have regard to deeds? Let me examine Mr. Gandhi's deeds.

The work which is claimed by Mr. Gandhi and his friends to have been done by him and the Congress for the Untouchables falls into two periods, the period which precedes the Poona Pact and the period which follows the Poona Pact. The first period may be called the period of the Bardoli Programme. The second period may be called the period of the Harijan Sevak Sangh.

First Period

To begin with the Bardoli Programme period. The Bardoli Programme or what is called the Constructive Programme of the Congress was the direct outcome of the new line of action adopted by the Congress in securing the political demands of the country. At the session of the Congress held at Nagpur in 1920 the Congress declared: Whereas the people of India are now determined to establish Swaraj; and Whereas all methods adopted by the people of India prior to the last special session of the Indian National Congress have failed to secure due recognition of their rights and liberties;

Now this Congress while reaffirming the resolution on nonviolent non-cooperation passed at the Special Session of the Congress at Calcutta declares that the entire or any part or parts of the scheme of non-violent non-cooperation, with the renunciation of voluntary association with the present Government at one end and the refusal to pay taxes at the other, should be put in force at a time to be determined by either the Indian National Congress or the All India Congress Committee and in the meanwhile to prepare the Country for it.

At the session of the Congress held at Ahmedabad in 1921 it was declared that:

> *"This Congress is further of opinion that Civil Disobedience is the only civilized and effective substitute for an armed rebellion.... and therefore advises all Congress Workers and others.... to organize individual civil disobedience and mass civil disobedience"..... It is to give effect to this policy of non-cooperation and civil disobedience and to prepare the people to take part in them that the Working Committee of the Congress met at Bardoli in February 1922 and drew up the following programme of action.*

"The Working Committee advises all Congress organisations to be engaged in the following activities:

(1) To enlist at least one crore of members of the Congress.

(2) To popularise the spinning wheel and to organise the manufacture of hand-spun and handwoven khaddar.

(3) To organise national schools.

(4) To organise the Depressed Classes for a better life, to improve their social, mental and moral condition to induce them to send their children to national schools and to provide for them the ordinary facilities which the other citizens enjoy.

Note: Whilst therefore where the prejudice against the Untouchables is still strong in places, separate schools and separate wells must be maintained out of Congress funds, every effort should be made to draw such children to national schools and to persuade the people to allow the Untouchables to use the common wells.

(5) To organise the temperance campaign amongst the people addicted to the drink habit by house-to-house visits and to rely more upon appeal to the drinker in his home than upon picketing.

(6) To organise village and town Panchayats for the private settlement of all disputes, reliance being placed solely upon the force of public opinion and the truthfulness of Panchayat decisions to ensure obedience to them.

(7) In order to promote and emphasize unity among all classes and races and mutual goodwill, the establishment of which is the aim of the movement of non-cooperation, to organise a social service department that will render help to all, irrespective of differences, in times of illness or accident.

(8) To continue the Tilak Memorial Swaraj Fund collections and call upon every Congressman or Congress sympathiser to pay at least a one-hundredth part of his annual income for 1921. Every province to send every month twenty-five per cent of its income from the Tilak Memorial Swaraj Fund to the All-India Congress Committee.

The above resolution shall be brought before the forthcoming session of the All-India Congress Committee for revision if necessary."

This programme was placed before the All-India Congress Committee at its meeting at Delhi on 20th February 1922 and was confirmed by the same. The programme is a very extensive programme and I am not concerned with what happened to the whole of it, how it was received and how it was worked out. I am concerned with only one item and that which relates to the Depressed Classes.

After it was confirmed by the All-India Congress Committee, the Working Committee met at Lucknow in June 1922 and passed the following resolution.

"This Committee hereby appoints a committee consisting of Swami Shraddhanandji, Mrs. Sarojini Naidu and Messrs. I. K. Yajnik and G. B. Deshpande to formulate a scheme embodying practical measures to be adopted for bettering the condition of the so-called Untouchables throughout the country and to place it for consideration before the next meeting of this Committee, the amount to be raised for the scheme to be Rs. 2 lacs for the present." This resolution was placed before the All-India Congress Committee at its meeting in Lucknow in June 1922. It accepted the resolution with the amendent that "the amount to be raised for the scheme should be 5 lacs for the present", instead of 2 lacks as put forth in the resolution of the Working Committee. How did this programme fare, what practical measures did the Committee suggest and how far were these measures given effect to? These questions one must ask in order to assess the work of Mr. Gandhi and the Congress for the Untouchables.

It seems that before the resolution appointing the Committee was adopted by the Working Committee, one of its Members Swami Shradhanand tendered his resignation of the membership of the Committee. For one finds that at the very sitting at which the Working Committee passed this resolution, another resolution to the following effect was passed by the Working Committee: "Read letter from Swami Shradhanandji, dated 8th June 1922 for an advance for drawing up a scheme for depressed classes work. Resolved that Mr. Gangadharrao B. Deshpande be appointed convener of the sub-committee appointed for the purpose and he be requested to convene a meeting at an early date, and that Swami Shradhanand's letter be referred to the sub-committee."

The Working Committee met again in July 1922 in Bombay and passed the following Resolution: "That the General Secretary be asked to request Swami Shradhanand to reconsider his resignation and withdraw it and a sum of Rs. 500/-be remitted to the Convener, Shri G. B. Deshpende, for the contingent expenses of the Depressed Classes Sub-Committee."

The year 1922 thus passed away without anything being done to further that item of the Bardoli Programme which related to the Depressed Classes. The year 1923 came on. The Working Committee met at Gaya in January 1923 and passed the following resolution: "With reference to Swami Shraddhanand's resignation, resolved that the remaining members of the Depressed Classes Subcommittee do form the Committee and Mr. Yajnik be the convener."

The All-India Congress Committee met in Feb. 1923 at Bombay and seeing that nothing was done as yet, recorded the following resolution: "Resolved that the question of the condition of the Untouchables be referred to the Working Committee for necessary action."

What did the Working Committee do then? It met at Poona on the 17th April 1923 and resolved as follows: "Resolved that while some improvement has been effected in the treatment of the so-called Untouchables in response to the policy of the Congress this Committee is conscious that much work remains yet to be done in this respect and inasmuchas this question of untouchability concerns the Hindu community particularly, it requests the All-India Hindu Mahasabha also to take up this matter and to make strenuous efforts to remove this evil from amidst the Hindu community."

Thus came to an end the Constructive Programme undertaken by Mr. Gandhi and the Congress for the Untouchables. The Bardoli programme for the Untouchables was in no sense a revolutionary programme. It did attempt to abolish untouchability. It does not attempt to break up caste. There is no mention of intermarriage or interdinning. It accepts the principle of separate wells and separate schools for Untouchables. It was purely an ameliorative programme. And yet such a harmless programme the Congress failed to carry through.

It must further be remembered that this was a time when the Congress was on the war path. It was determined to fight British Imperialism and was most anxious to draw every community towards itself and make all disaffected towards the British. This was the time when the Congress could have been expected to show to the Untouchables that the Congress stood for them and was prepared to serve them in the same way that it was prepared to serve the Musalmans. There could be no more propitious circumstance which could make the Hindus overcome their antipaty towards the Untouchables and undertake to serve. But even such propitious circumstance did not prove sufficient to energize Congressmen to do this small bit for the Untouchables. How hard must be the anti-social feelings of the Hindus against the Untouchables that even the highest bliss and the greatest stimulant, namely the prospect of winning Swaraj, were not sufficient to dissolve that spirit. The tragedy and the shamelessness of this failure by the Congress to carry through their programme for the Untouchables is aggravated by the way in which the matter was disposed of.

The work of the amelioration of the Untouchables could not have been left in worse hands. If there is any body which is quite unfit for addressing itself to the problem of the Untouchables, it is the Hindu Mahasabha. It is a militant Hindu organization. Its aim and object is to conserve in every way everything that is Hindu, religious and culture. It is not a social reform association. It is a purely political organization whose main object and aim is to combat the influence of the Muslims in Indian politics. Just to preserve its political strength it wants to maintain its social solidarity and its way to maintain social solidarity is not to talk about caste or untouchability. How such a body could have been selected by the Congress for carrying on the work of the Untouchables passes my comprehension. This shows that the Congress wanted somehow to get rid of an inconvenient problem and wash its hands of it.

The Hindu Mahasabha, of course, did not come forth to undertake the work and the Congress had merely passed a pious resolution recommending the work to them without making any promise for financial provision. So the project came to an inglorious and ignominious end. Yet there will not be wanting thousands of Congressmen who would not be ashamed to boast that the Congress has been fighting for the cause of the Untouchables and what is worse is that there will not be wanting hundreds of foreigners who are ready to believe it under the false propaganda carried on by men like Charles F. Andrews, who is the friend of Mr. Gandhi and who thinks that to popularize Gandhi in the Western World is his real mission in life.'

It is not enough to know that the effort failed and had to be wound up. It is necessary to inquire why Swami Shradhanand resigned and refused to serve on the proposed Committee. There must be some good reason for it. For the Swami was the most enlightened Arya Samajist and very conscientiously believed in the removal of untouchability. On this point, the correspondence that passed between the Swami and the General Secretary to the All India Congress Committee throws a flood of light on the mentality of the Congressmen and I make no apology for reproducing below the whole of it.

Draft of Gandhi-Muslim Pact

Muslim Delegation to the Round Table Conference

The following proposals were discussed by Mr. Gandhi and the Muslim Delegation at 10 p.m. last night. They are divided into two parts—The proposals made by the Muslims for safeguarding their rights and the proposals made by Mr. Gandhi regarding the Congress policy. They are given herewith as approved by Mr. Gandhi, and placed for submission to the Muslim Delegation for their opinion.

Muslim Proposals *Proposal*	*Gandhi's*
1. In the Punjab and Bengal bare majority of one percent of Musalmans but the question of whether it should be by means of joint electorates and reservation of 51 per cent of the whole house should be referred to the Musalman voters before the new constitution comes into force and their verdict should be accepted.	1. That the Franchise should be on the basis of adult suffrage. 2. *No special reservations to any other community save Sikhs and Hindu Minorities.*

2. In other provinces where the Musalmans are in a minority the present weightage enjoyed by them to continue, but whether the seats should be reserved to a joint electorate, or whether they should have separate electorates should be determined by the Musalman voters by a referendum under the new constitution, and their verdict should be accepted.
3. That the Musalman representatives to the Central Legislature in both the houses should be 26 percent of the total number of the British India representatives, and 7 percent at least by convention should be Musalmans, out of the quota that may be assigned to Indian States, that is to say, one-third of the whole house when taken together.
4. That the residuary power should vest in the federating Provinces of British India.
5. That the other points as follows being agreed to:
   1. Sindh.
   2. N.W.F.P.
   3. Services
   4. Cabinet
   5. Fundamental rights and safeguards for religion and culture.
   6. Safeguards against legislation affecting any community.

The Congress demands:

A. Complete Independence

B. Complete control over the defence immediately.

C. Complete control over external affairs.

D. Complete control over finance.

E. Investigation of public debts and other obligations by an independent tribunal.

F. As in the case of a partnership, right of either party to terminate it.

This is the agreement which Mr. Gandhi was prepared to enter with the Musalmans. By this agreement Mr. Gandhi was prepared to give to the Musalmans the fourteen points they had been demanding. In return Mr. Gandhi wanted the Musalmans among other things to agree to continue the benefit of the principle of special representation to Hindus, Muslims and Sikhs. Some one might ask what is wrong in such an agreement. Has not the Congress said that they will not agree to extend communal representation to others besides these three? Such a view cannot but be treated as a superficial view. Those who see nothing wrong in it must answer two questions. First is this. Where was the necessity for Mr. Gandhi to get the Musalmans to agree to the Congress policy of not extending the benefit of special representation to other minorities and the untouchables. Mr. Gandhi could have said as the Congress had been saying to the other minorities he was not prepared to agree to their claim. Why did he want the Musalmans to join him in resisting their claim? And if this was not his object why did he make it a term of the agreement which the Musalmans were to perform in return for what he agreed to do for them.

Secondly why did Mr. Gandhi come forward to give the Musalmans their fourteen demands at this particular juncture. These fourteen political demands of the Musalmans rightly or wrongly were rejected by all. They were rejected by the Hindu Maha-Sabha. They were rejected by the Simon Commission. They were rejected by the Congress. There was no support for these 14 demands of the Musalmans from any quarter whatsoever. Why did Mr. Gandhi become ready to grant them except with the object of buying the Musalmans so that with their help he could more effectively resist the demand of the other minorities and the untouchables?

In my view Mr. Gandhi was not engaged in making any bona-fide agreement. He was inducing the Musalmans to join in a conspiracy with him to resist the claim of the smaller minorities and the untouchables. It was not an agreement with the Musalmans. It was a plot against the Untouchables. It was worse, it was a stab in the back.

This so-called agreement fell through because among other reasons it was impossible for the Mohammedans to agree to the exclusion of the Untouchables from the benefit of special representation. How could the Muslims agree to such a project? They were fighting for special representation for Muslims. They were not only fighting for special representation, they were fighting for weightage in representation. They knew that the case for Muslims rested only on the ground that India was once ruled by the Musalmans, that they had political importance to maintain and as Hindus are likely to discriminate against Muslims in elections to the Legislatures, there may not be sufficient Muslims returned to the Legislature, that the Muslims will sink politically and that to prevent such a calamity they must be given special representation. As against this one ground in favour of Muslims there were a hundred grounds in favour of the claim by the Untouchables. With what face could the Musalmans oppose this demand of the Untouchables?

The Musalmans had not lost their balance or their sense of shame. They refused to be party to such a deal—a deal which they could not publicly defend. Mr. Gandhi still kept on pestering the Musalmans. When he could not induce them to accept the price he offered namely the grant of fourteen points—because they felt that the world would not call it price but would call it the wages of sin, Mr. Gandhi sought to appeal to the religious scruples of the Musalmans. The day before the 13th November, 1931 when the minorities pact was presented to the Minorities Sub-Committee of the Round Table Conference Mr. Gandhi took a copy of the Koran and went to the Ritz Hotel in Piccadilly where the Rt. Hon. H. H. Aga Khan was staying to meet the Muslim delegates who had assembled there. To Muslim delegates he asked—" Why are you dividing the Hindu Community which you are doing by recognising the claim of the Untouchables for separate representation? Does the Koran sanction such a deed? Show me where it does? If you cannot, will you not stop perpetrating such a crime upon your sister Community?" I do not know how the Muslim delegates answered this question of Mr. Gandhi. It must have been a very difficult question for them to answer. Such a contingency could not have been present to the mind of the Holy Prophet and he could not have provided for it specifically. His followers knew that contingencies would arise for which he had given no directions and they had therefore asked him what they should do, and the Prophet had given them this general direction. He said to them, "in such a case see what the Kaffirs are doing and do just the opposite of it". Whether the Muslim delegates relied upon this to answer Mr. Gandhi is more than I can say. What I have stated is what I have heard and my source is the most authentic source. Here again Mr. Gandhi failed because the next day in the open Committee when Mr. Gandhi let loose his fury against the Untouchables, the Mohammedans were silent.

What can one say of this conduct of Mr. Gandhi? Mr. Bernard Shaw has said that the British do everything on principle. Similarly, Mr. Gandhi says he does everything on the principle of morality and good faith. Can the acts of Mr. Gandhi be justified by tests of justice and good faith? I wonder. Let me state a few facts.

Before I left for London for the first Round Table Conference I had met Mr. Gandhi in Bombay. At that meeting I had informed Mr. Gandhi that at the Round Table Conference I would be asking for special representation for the Untouchables. Mr. Gandhi would not consent. But he also told me that he would not oppose. I felt that it was just a case of difference of opinion. At the second Round Table Conference I met Mr. Gandhi twice, once alone and second time along with the representatives of the smaller minorities. At the first meeting Mr. Gandhi was spinning and I was talking. I spoke for an hour during the whole of which he did not utter even a word. At the end he just said this much. I have now heard you. I will think over what you have said.' At the second interview he again heard me and the representatives of the smaller minorities and he told me that he was not prepared to agree to the claim I was making on behalf of the Untouchables. Thereafter the Minorities Sub-Committee was convened on 28th September 1931. At the meeting of the Sub-Committee on 1st October 1931, the following motion was made by Mr. Gandhi:

> *"Prime Minister, after consultation with His Highness the Aga Khan and other Muslim friends last night, we came to the conclusion that the purpose for which we meet here would be better served if a week's adjournment was asked for. I have not had the opportunity of consulting my other colleagues, but I have no doubt that they will also agree in the proposal I am making."*

The proposal was seconded by the Aga Khan. I at once got up and objected to the motion and in support of my objection made the following statement:

> *I do not wish to create any difficulty in our making every possible attempt to arrive at some solution of the problem with which this Committee has to deal, and if a solution can be arrived at by the means suggested by Mahatma Gandhi, I, for one, will have no objection to that proposal.*

But there is just this one difficulty with which I, as representing the Depressed Classes, am faced. I do not know what sort of committee Mahatma Gandhi proposes to appoint to consider this question during the period of adjournment, but I suppose that the Depressed Classes will be represented on this Committee. *Mr. Gandhi:* Without doubt.

Thank you. But I do not know whether in the position in which I am today it would be of any use for me to work on the proposed Committee. And for this reason. Mahatma Gandhi told us on the first day that he spoke in the Federal Structure Committee that as a representative of the Indian National Congress he was not prepared to give political recognition to any community other than the Muhammadans and the Sikhs. He was not prepared to recognize the Anglo-indians, the Depressed Classes, and the Indian Christians. I do not think that I am doing any violence to etiquette by stating in this Committee that when I had the pleasure of meeting Mahatma Gandhi a week ago and discussing the question of the Depressed Classes with him, and when we, as members of the other minorities, had the chance of talking with him yesterday, in his office, he told us in quite plain terms that the attitude that he had taken in the Federal Structure Committee was a firm and well considered attitude.

What I would like to say is that unless at the outset I know that the Depressed Classes are going to be recognised as a community entitled to political recognition in the future Constitution of India, I do not know whether it will serve any purpose for me to join the committee that is proposed by Mahatma Gandhi to be constituted to go into this matter. Unless, therefore, I have an assurance that this Committee will start with the assumption that all those communities which the Minorities Subcommittee last year recommended as fit for recognition in the future constitution of India will be included, I do not know that I can whole-heartedly support the proposition for adjournment, or that I can whole-heartedly cooperate with the Committee that is going to be nominated. That is what I wish to be clear about.

I should like to make my position further clear. It seems that there has been a certain misunderstanding regarding what I said. It is not that I object to adjournment; it is not that

I object to serving on any Committee that might be appointed to consider the question. What I would like to know before I enter upon this Committee, if they give me the privilege of serving on it, is: What is the thing that this Committee is going to consider? Is it only going to consider the question of the Muhammadans *vis-a-vis* the Hindus? Is it going to consider the question of the Muhammadans *vis-a-vis* the Sikhs in the Punjab? Or is it going to consider the question of the Christians, the Anglo-indians and the Depressed Classes?

"If we understand perfectly well before we start that this committee will not merely concern itself with the question of the Hindus and the Muhammadans, of the Hindus and the Sikhs, but will also take upon itself the responsibility of considering the case of the Depressed Classes, the Anglo-indians and the Christians, I am perfectly willing to allow this adjournment resolution to be passed without any objection. But I do want to say this, that if I am to be left out in the cold and if this interval is going to be utilised for the purposes of solving the Hindu-Muslim question, I would press that the Minorities Committee should itself grapple with the question and consider it, rather than allow the question to be dealt with by some other informal Committee for arriving at a solution of the communal question in respect of some minorities only."

The Prime Minister as Chairman of the Committee called upon Mr. Gandhi to explain his position and Mr. Gandhi made the following statement in reply: "Prime Minister and friends, I see that there is some kind of misunderstanding with reference to the scope of the work that some of us have set before ourselves. I fear that Dr. Ambedkar, Colonel Gidney and other friends are unnecessarily nervous about what is going to happen. Who am I to deny political status to any single interest or class or even individual in India?

As a representative of the Congress I should be unworthy of the trust that has been reposed in me by the Congress if I were guilty of sacrificing a single national interest. I have undoubtedly given expression to my own views on these points. I must confess that I hold to those views also. But there are ways and ways of guaranteeing protection to every single interest. It will be for those of us who will be putting our heads together to try to evolve a scheme. Nobody would be hampered in pressing his own views on the members of this very informal conference or meeting.

> *"I do not think, therefore, that anybody need be afraid as to being able to express his opinion or carrying his opinion also. Mine will be there equal to that of every one of us; it will carry no greater weight; I have no authority behind me to carry my opinion against the opinion of anybody. I have simply given expression to my views in the national interest, and I shall give expression to these views whenever they are opportune. It will be for you, it is for you to reject or accept these opinions. Therefore please disburse your minds, to everyone of us, of the idea that there is going to be any steam-rolling in the Conference and the informal meetings that I have adumbrated. But if you think that this is one way of coming closer together than by sitting stiffly at this table, you will not carry this adjournment motion but give your whole-hearted cooperation to the proposal that I have made in connection with these informal meetings."*

I then withdrew my objection.

Now here is a definite word given by Mr. Gandhi in open Conference—namely that if all others agreed to recognize the claim of the Untouchables he would not object. And after having given this word Mr. Gandhi went about inducing the Musalmans not to recognize the claim of the Untouchables and to bribe them to resile and take back their plighted word! Is this good faith or is this treachery? If this is not treachery I wonder what else could be called treachery.

I was pilloried because I signed what is called the Minorities Pact. I was depicted as a traitor. I have never been ashamed of my signature to the pact. I only pity the ignorance of my critics. They forget that the minorities could have taken the same attitude that Ulster took towards Irish Home Rule.

Redmond was prepared to offer any safeguard to Ulstermen. The Ulstermen's reply was, "Damn your safeguards we don't wish to be ruled by you". The Hindus ought to thank the minorities that they did not take any such attitude. All the Pact contained were Safeguards and nothing more. Instead of thanking them Mr. Gandhi poured his vials of wrath upon the pact and its authors. He said:

> *"Coming to this document', I accept the thanks that have been given to me by Sir Hubert Carr. Had it not been for the remarks that I made when I shouldered that burden, and had it not been for my utter failure to bring about a solution. Sir Hubert Carr rightly says he would not have found the very admirable solution that he has been able, in common with the other minorities, to present to this Committee for consideration and finally for the consideration and approval of His Majesty's Government."*

"I will not deprive Sir Hubert Carr and his associates of the feeling of satisfaction that evidently actuates them, but, in my opinion, what they have done is to sit by the carcass, and they have performed the laudable feat of dissecting that carcass."

1 Reference is to the Minorities Pact.

Had Mr. Gandhi any right to be indignant? Had he any right to feel morally offended? Was he entitled to throw stones at the Minorities? Mr. Gandhi forgot that he was as much a sinner as the Minorities and worse he was a sinner without a sense of justice. For if the Minorities were dividing the carcass what was Mr. Gandhi himself doing? He too was busy in dividing the carcass. The only difference between Mr. Gandhi and the Minorities was this—Mr. Gandhi wanted that the carcass should be divided among three only, Hindus, Musalmans and Sikhs. The Minorities wanted that others also should be given a share, and which of these two can claim to have justice, on its side, Mr. Gandhi who wanted that the division of the carcass should be to strong sturdy well-nourished wolves or the minorities who pressed that the lean and hungry lambs should also be given a morsel? Surely in this controversy justice was not on the side of Mr. Gandhi.

Mr. Gandhi was the man who claimed to be the Champion of the Untouchables better than those who belonged to the Untouchables themselves. Claiming to be their champion he refused without any regard to morality, justice and necessity, their claim to representation which could be their only way to protection against social tyranny and social oppression while he was prepared to give to the Musalmans, the Hindus and the Sikhs a goodly share of political power. When others far better placed were claiming for power, Mr. Gandhi wanted the Untouchables to live under his providence and that of the Congress without any means of protection knowing full well that their lives were exposed to danger and humiliation every moment and when he came to know that the Untouchables were seeking outside aid in support of their claim Mr. Gandhi resorted to a terrible act of treachery. Were the Untouchables unjustified in presenting to the world their charge sheet against Mr. Gandhi when he arrived in Bombay from the Round Table Conference?

Gandhi and his Fast

The Communal question was the rock on which the Indian Round Table Conference suffered a shipwreck. The Conference broke up as there could be no agreement between the majority and minority communities. The minorities in India insisted that their position under Swaraj should be safeguarded by allowing them special representation in the Legislatures. Mr. Gandhi as representative of the Congress was not prepared to recognize such a claim except in the case of the Muslims and the Sikhs. Even in the case of the Muslims and Sikhs, no agreement was reached either on the question of the number of seats or the nature of the electorates.

There was a complete deadlock. As there was no possibility of an agreement, the hope lay in arbitration. On this everybody was agreed except myself and it was left to Mr. Ramsay Macdonald, the Prime Minister to decide upon the issue.

When at the first Round Table Conference, the Indian delegates did not agree upon a solution of the Communal question, followers of Mr. Gandhi said that nothing better could be expected from them. It was said that they were unrepresentative and responsible to nobody and were deliberately creating disunity by playing into the hands of the British whose tools and nominees they were. The world was told to await the arrival of Mr. Gandhi, whose statesmanship it was promised would be quite adequate to settle the dispute. It was therefore a matter of great humiliation for the friends of Mr. Gandhi that he should have acknowledged his bankruptcy and joined in the request to the Prime Minister to arbitrate.

But if the Conference failed the fault is entirely of Mr. Gandhi. A more ignorant and more tactless representative could not have been sent to a Conference which was convened to forge a constitution which was to reconcile the diverse interests of India. Mr. Gandhi was thoroughly ignorant of Constitutional Law or Finance. He does not believe in intellectual equipment. Indeed he has a supreme contempt for it and his contributions to the solutions of the many difficulties is therefore nil. He was tactless because he annoyed almost all the delegates by constantly telling them that they were nonentities and he was the only man who counted and who could deliver the goods. At the first Round Table Conference the delegates did not agree upon a solution of the communal problem. But it is equally true that they were very near agreeing to it and when they departed they had not given up hope of agreeing. But at the end of the second Round Table Conference, so much bad blood was created by Mr. Gandhi that there was no chance of reconciliation left and there was no way except arbitration. The Prime Minister's decision on the communal question was announced on 17th August 1932. The terms of the decision in so far as they related to the Untouchables were as follows:

Communal Decision by His Majesty's Government 1932

1. In the statement made by the Prime Minister on 1st December last on behalf of His Majesty's Government at the close of the second session of the Round Table Conference, which was immediately afterwards endorsed by both Houses of Parliament, it was made plain that if the communities in India were unable to reach a settlement acceptable to all parties on the communal questions which the Conference had failed to solve. His Majesty's Government were determined that India's constitutional advance should not on that account be frustrated, and that they would remove this obstacle by divising and applying themselves a provisional scheme.

2. On the 19th March last His Majesty's Government, having been informed that the continued failure of the communities to reach agreement was blocking the progress of the plans for the framing of a new Constitution, stated that they were engaged upon a careful re-examination of the difficult and controversial question which arise. They are now satisfied that without a decision of at least some aspects of the problems connected with the position of minorities under the new Constitution, no further progress can be made with the framing of the Constitution.
3. His Majesty's Government have accordingly decided that they will include provisions to give effect to the scheme set out below in the proposals relating to the Indian Constitution to be laid in due course before Parliament. The scope of this scheme is purposely confined to the arrangements to be made for the representation of the British Indian communities in the Provincial Legislatures, consideration of representation in the Legislature at the Centre being deferred for the reason given in paragraph 20 below. The decision to limit the scope of the scheme implies no failure to realize that the framing of the Constitution will necessitate the decision of a number of other problems of great importance to minorities, but has been taken in the hope that once a pronouncement has been made upon the basic questions of method and proportions of representation the communities themselves may find it possible to arrive at *modus vivendi* on other communal problems, which have not received the examination they require.
4. His Majesty's Government wish it to be most clearly understood that they themselves can be no parties to any negotiations which may be initiated with a view to the revision of their decision, and will not be prepared to give consideration to any representation aimed at securing the modification of it which is not supported by all the parties affected. But they are most desirous to close no door to an agreed settlement should such happily be forthcoming. If, therefore, before a new

Government of India Act has passed into law, they are satisfied that the communities who are concerned are mutually agreed upon a practicable alternative scheme, either in respect of any one or more of the Governors' Provinces or in respect of the whole of the British India, they will be prepared to recommend to Parliament that alternative should be substituted for the provisions now outlined.

5. Members of the "depressed classes" qualified to vote will vote in a general constituency. In view of the fact that for a considerable period these classes would be unlikely, by this means alone, to secure any adequate representation in the Legislature, a number of special seats will be assigned to them as shown in the table. These seats will be filled by election from special constituencies in which only members of the "depressed classes" electorally qualified will be entitled to vote. Any person voting in such a special constituency will, as stated above, be also entitled to vote in a general costituency. It is intended that these constituencies should be formed in selected areas where the depressed classes are most numerous, and that, except in Madras, they should not cover the whole area of the Province.

In Bengal it seems possible that in some general constituencies a majority of the voters will belong to the Depressed Classes. Accordingly, pending further investigation no number has been fixed for the members to be returned from the special Depressed Class constituencies in that Province. It is intended to secure that the Depressed, Classes should obtain not less than 10 seats in the Bengal Legislature.

The precise definition in each Province of those who (if electorally qualified) will be entitled to vote in the special Depressed Class constituencies has not yet been finally determined. It will be based as a rule on the general principles advocated in the Franchise Committee's Report. Modification may, however, be found necessary in some Provinces in Northern India where the application of the general criteria of untouchability might result in a definition unsuitable in some respects to the special conditions of the Province. His Majesty's Government do not consider that these special Depressed

Classes constituencies will be required for more than limited time. They intend that the Constitution shall provide that they shall come to an end after 20 years if they have not previously been abolished under the general powers of electoral revision referred to in paragraph 6.

So far as the other minority communities were concerned the Communal Award was accepted and the sore of disunity and discord was closed. But so far as the Untouchables were concerned it remained open. Mr. Gandhi would not allow it to be healed. On his return to India from the Round Table Conference Mr. Gandhi was put behind the bars by the British Government. But though in the Yeravada gaol Mr. Gandhi had not forgotten that he had to prevent the Untouchables from getting their claim to special representation recognized by the British Government. He feared that the British Government might grant them this right notwithstanding the threat he had held out while at the Round Table Conference to resist it with his own life. Consequently he took the earliest opportunity to be in communication with the very British Government which had incarcerated him.

On the 11th March 1932 Mr. Gandhi addressed the following letter to Sir Samuel Hoare, the then Secretary of State for India: Dear Sir Samuel,You will perhaps recollect that at the end of my speech at the Round Table Conference when the minorities' claim was presented, I had said that I should resist with my life the grant of separate electorates to the Depressed Classes.

This was not said in the heat of the moment nor by way of rhetoric. It was meant to be a serious statement. In pursuance of that statement I had hoped on my return to India to mobilize public opinion against separate electorate, at any rate for the Depressed Classes. But it was not to be.

From the newspapers I am permitted to read, I observe that any moment His Majesty's Government may declare their decision. At first I had thought, if the decision was found to create separate electorates for the Depressed Classes, I should take such steps as I might then consider necessary to give effect to my vow. But I feel it would be unfair to the British Government for me to act without giving previous notice.

Naturally, they could not attach the significance I give to my statement.

Separate Electorates Harmful: I need hardly reiterate all the objections I have to the creation of separate electorates for the Depressed Classes. I feel as if I was one of them. Their case stands on a wholly different footing from that of others. I am not against their representation in the legislatures. I should favour every one of their adults, male and female, being registered as voters irrespective of education or property qualification, even though the franchise test may be stricter for others. But I hold that separate electorates is harmful for them and for Hinduism, whatever it may be from the purely political standpoint. To appreciate the harm that separate electorates would do them one has to know how they are distributed amongst the so-called Caste Hindus and how dependent they are on the latter. So far as Hinduism is concerned, separate electorate would simply vivisect and disrupt it.

For me the question of these classes is predominantly moral and religious. The political aspect, important though it is, dwindles into significance compared to the moral and religious issue.

You will have to appreciate my feelings in this matter by remebering that I have been interested in the condition of these classes from my boyhood and have more than once staked my all for their sake. I say this not to pride myself in any way. For, I feel that no penance that the Hindus may do can in any way compensate for the calculated degradation to which they have consigned the Depressed Classes for centuries.

"Shall fast unto Death"

But I know that separate electorate is neither a penance nor any remedy for the crushing degradation they have groaned under. I, therefore, respectfully inform His Majesty's Government that in the event of their decision creating separate electorate for the Depressed Classes, I must fast unto death.

I am painfully conscious of the fact that such a step whilst I am a prisoner, must cause grave embarrassment to His Majesty's Government, and that it will be regarded by many as highly improper on the part of one holding my position to introduce into the political field methods which they would describe as hysterical if not much worse. All I can urge in defence is that for me the contemplated step is not a method, it is part of my being. It is the call of conscience which I dare not disobey, even though it may cost whatever reputation for sanity I may possess. So far as I can see now, my discharge from imprisonment would not make the duty of fasting any the less imperative. I am hoping, however, all my fears are wholly unjustified and the British Government have no intention whatever of creating separate electorate for the Depressed Classes.

The following reply was sent to Mr. Gandhi by the Secretary of State: There can be no doubt that Sir Samuel Hoare has showed you and the Cabinet my letter to him of 11th March on the question of the representation of the Depressed Classes. That letter should be treated as part of this letter and be read together with this.

Decision to Fast

I have read the British Government's decision on the representation of Minorities and have slept over it. In pursuance of my letter to Sir Samuel Hoare and my declaration at the meeting of the Minorities Committee of the Round Table Conference on 13th November, 1931, at St. James' Palace, I have to resist your decision with my life. The only way I can do so is by declaring a perpetual fast unto death from food of any kind save water with or without salt and soda. This fast will cease if during its progress the British Government, of its own motion or under pressure of public opinion, revise their decision and withdraw their scheme of communal electorates for the Depressed Classes, whose representatives should be elected by the general electorate under the common franchise no matter how wide it is.

The proposed fast will come into operation in the ordinary course from the noon of 20th September next, unless the said decision is meanwhile revised in the manner suggested above.

I am asking the authorities here to cable the text of this letter to you so as to give you ample notice. But in any case, I am leaving sufficient time for this letter to reach you in time by the slowest route.

I also ask that this letter and my letter to Sir Samuel Hoare already referred to be published at the earliest possible moment. On my part, I have scrupulously observed the rule of the jail and have communicated my desire or the contents of the two letters to no one, save my two companions, Sardar Vallabhabhai Patel and Mr. Mahadev Desai. But I want, if you make it possible, public opinion to be affected by my letters. Hence my request for their early publication.

Not to Compass Release

I regret the decision I have taken. But as a man of religion that I hold myself to be, I have no other course left open to me. As I have said in my letter to Sir Samuel Hoare, even if His Majesty's Government decided to release me in order to save themselves from embarrassment, my fast will have to continue. For, I cannot now hope to resist the decision by any other means; And I have no desire whatsoever to compass my release by any means other than honourable.

It may be that my judgement is warped and that I am wholly in error in regarding separate electorates for the Depressed Classes as harmful to them or to Hinduism. If so, I am not likely to be in the right with reference to other parts of my philosophy of life. In that case my death by fasting will be at once a penance for my error and a lifting of a weight from off these numberless men and women who have childlike faith in my wisdom. Whereas if my judgement is right, as I have little doubt it is, the contemplated step is but due to the fulfilment of the scheme of life which I have tried for more than a quarter of a century, apparently not without considerable success.

The Prime Minister replied as under:

I have received your letter with much surprise and, let me add, with very sincere regret. Moreover, I cannot help thinking that you have written it under a misunderstanding as to what the decision of His Majesty's Government as regards the Depressed Classes really implies. We have always understood you were irrevocably opposed to the permanent segregation of the Depressed Classes from the Hindu community. You made your position very clear on the Minorities Committee of the

Round Table Conference and you expressed it again in the letter you wrote to Sir Samuel Hoare on 11th March. We also knew your view was shared by the great body of Hindu opinion, and we, therefore, took it into most careful account when we were considering the question of representation of the Depressed Classes.

Government Decision Explained: Whilst, in view of the numerous appeals we have received from Depressed Class organisations and the generally admitted social disabilities under which they labour and which you have often recognized, we felt it our duty to safeguard what we believed to be the right of the Depressed Classes to a fair proportion of representation in the legislatures, we were equally careful to do nothing that would split off their community from the Hindu world. You yourself stated in your letter of March 11, that you were not against their representation in the legislatures.

Under the Government scheme the Depressed Classes will remain part of the Hindu community and will vote with the Hindu electorate on an equal footing, but for the first twenty years, while still remaining electorally part of the Hindu community, they will receive through a limited number of special constituencies, means of safeguarding their rights and interests that, we are convinced, is necessary under present conditions.

Where these constituencies are created, members of the Depressed Classes will not be deprived of their votes in the general Hindu constituencies, but will have two votes in order that their membership of the Hindu community should remain unimpaired.

We have deliberately decided against the creation of what you describe as a communal electorate for the Depressed Classes and included all Depressed Class voters in the general or Hindu constituencies so that the higher caste candidates should have to solicit their votes or Depressed Class candidates should have to solicit the votes of the higher castes at elections. Thus in every way was the unity of Hindu society preserved.

Safeguard Temporary: We felt, however, that during the early period of responsible government when power in the Provinces would pass to whoever possessed a majority in the legislatures, it was essential that the Depressed Classes, whom you have yourself described in your letter to Sir Samuel Hoare as having consigned by Caste Hindus to calculated degradation for centuries, should return a certain number of members of their own choosing to legislatures of seven of the nine provinces to voice their grievances and their ideals and prevent decisions going against them without the legislature and the Government listening to their case — in a word, to place them in a position to speak for themselves which every fair-minded person must agree to be necessary.

We did not consider the method of electing special representatives by reservation of seats in the existing conditions, under any system of franchise which is practicable, members who could genuinely represent them and be responsible for them, because in practically all cases, such members would be elected by a majority consisting of higher caste Hindus.

The special advantage initially given under our scheme to the Depressed Classes by means of a limited number of special constituencies in addition to their normal electoral rights in the general Hindu constituencies is wholly different in conception and effect from the method of representation adopted for a minority such as the Moslems by means of separate communal electorates.

For example, a Moslem cannot vote or be a candidate in a general constituency, whereas any electorally qualified member of the Depressed Classes can vote in and stand for the general constituency.

Reservation Minimum: The number of territorial seats allotted to Moslems is naturally conditioned by the fact that it is impossible for them to gain any further territorial seats and in most provinces they enjoy weightage in excess of their population ratio; the number of special seats to be filled from special Depressed Classes constituencies will be seen to be small and has been fixed not to provide a quota numerically appropriate for the total representation of the whole of the Depressed Class population, but solely to secure a minimum number of spokesmen for the Depressed Classes in the legislature who are chosen exclusively by the Depressed Classes. The proportion of their special seats is everywhere much below the population percentage of the Depressed Classes. As I understand your attitude, you propose to adopt the extreme course of starving yourself to death not in order to secure that the Depressed Classes should have joint electorates with other Hindus, because that it already provided, nor to maintain the unity of Hindus, which is also provided, but solely to prevent the Depressed Classes, who admittedly suffer from terrible disabilities today, from being able to secure a limited number of representatives of their own choosing to speak on their behalf in the legislatures which will have a dominating influence over their future.

In the light of these very fair and cautious proposals, I am quite unable to understand the reason of the decision you have taken and can only think you have made it under a misapprehension of the actual facts.

Government Decision Stands: In response to a very general request from Indians after they had failed to produce a settlement themselves the Government much against its will, undertook to give a decision on the minorities question. They have now given it, and they cannot be expected to alter it except on the conditions they have stated.

I am afraid, therefore, that my answer to you must be that the Government's decision stands and that only agreement of the communities themselves can substitute other electoral arrangements for those that Government have devised in a sincere endeavour to weigh the conflicting claims on their just merits.

You ask that this correspondence, including your letter to Sir Samuel Hoare of March 11th, should be published. As it would seem to me unfair if your present internment were to deprive you of the opportunity of explaining to the public the reason why you intend to fast, I readily accede to the request if on reconsideration you repeat it. Let me, however, once again urge you to consider the actual details of Government's decision and ask yourself seriously the question whether it really justifies you in taking the action you contemplate.

Finding that the Prime Minister would not yield he sent him the following letter informing him that he was determined to carry out his threat of fast unto death.

I have to thank you for your frank and full letter telegraphed and received this day. I am sorry, however, that you put upon the contemplated step an interpretation that never crossed my mind. I have claimed to speak on behalf of the very class, to sacrifice whose interests you impute to me a desire to fast myself to death. I had hoped that the extreme step itself would effectively prevent any such selfish interpretation without arguing, I affirm that for me this matter is one of pure religion. The mere fact of the Depressed Classes having double votes does not protect them or Hindu society in general from being disrupted.

In the establishment of separate electorate at all for the Depressed Classes I sense the injection of poison that is calculated to destroy Hinduism and do no good whatever to the Depressed Classes. You will please permit me to say that no matter how sympathetic you may be, you cannot come to a correct decision on a matter of such vital and religious importance to the parties concerned.

I should not be against even over-representation of the Depressed Classes. What I am against is their statutory separation even in a limited form, from the Hindu fold, so long as they choose to belong to it. Do you realize that if your decision stands and the constitution comes into being, you arrest the marvellous growth of the work of Hindu reformers who have dedicated themselves to the uplift of their suppressed brethren in every walk of life?

Decision Unchanged

I have, therefore, been compelled reluctantly to adhere to the decision conveyed to you. As your letter may give rise to a misunderstanding, I wish to state that the fact of my having isolated for special treatment the Depressed Classes question from other parts of your decision does not in any way mean that I approve of or am reconciled to other parts of the decision. In my opinion, many other parts are open to very grave objection. Only I do not consider them to be any warrant for calling from me such self immolation as my conscience has prompted me to, in the matter of the Depressed Classes.

Accordingly on the 20th September 1932 Mr. Gandhi commenced his "fast unto death" as a protest against the grant of separate electorates to the Untouchables. The story of this fast has been told by Mr. Pyarelal in a volume which bears the picturesque and flamboyant title of "The Epic Fast". In the pages of this *lours Boswelliana* the curious will find all he wants to know about the happenings in India during these mad days and I need say nothing about it here. Suffice it to say that although Mr. Gandhi went on fast unto death he did not want to die. He very much wanted to live.

The fast therefore created a problem and that problem was how to save Mr. Gandhi's life. The only way to save his life was to alter the Communal Award so as not to hurt Mr. Gandhi's conscience.

The Prime Minister had made it quite clear that the British Cabinet would not withdraw it or alter it of its own but that they were ready to substitute for it a formula that may be agreed upon by the Caste Hindus and the Untouchables. As I had the privilege of representing the Untouchables at the Round Table Conference it was assumed that the assent of the Untouchables would not be valid unless I was a party to it.

At the moment my position as the representative of the Untouchables of India was not only not questioned but was accepted as a fact. All eyes naturally turned to me as the man or rather as the villain of the piece. Mr. Gandhi's life as he himself said was in my hands.

It is no exaggeration to say that no man was placed in a greater and graver dilemma than I was then. It was a baffling situation. I had to make a choice between two different alternatives.

There was before me the duty which I owed as a part of common humanity to save Gandhi from sure death. There was before me the problem of saving for the Untouchables the political rights which the Prime Minister had given them. I responded to the call of humanity and saved the life of Mr. Gandhi by agreeing to alter the Communal Award in a manner satisfactory to Mr. Gandhi. This agreement is known as the Poona Pact. The terms of the Poona Pact were as under:

1. There shall be seats reserved for the Depressed Classes out of general electorates. Seats in Provincial Legislatures shall be as follows:

Madras	30
Bombay with Sindh	15
Punjab	8
Bihar and Orissa	18
Central Provinces	20
Assam	7
Bengal	30
United Provinces	20
Total	148

This number of 148 seats was raised to 151 in making adjustments on seats for Bihar and Orissa. These figures are based on the total strength of the Provincial Councils announced in the Prime Minister's decision.

2. Election to these seats shall be by joint electorates subject, however, to the following procedure: All members of the Depressed Classes registered in the general electoral roll of a constituency, will form an electoral college which will elect a panel of four candidates belonging to the Depressed Classes, for each of such reserved seats by the method of single vote and four persons getting the highest number of votes in such primary election, shall be the candidates for election by the general electorates.
3. Representation of the Depressed Classes in the Central Legislature shall likewise be on the principle of joint electorates and reserved seats by the method of primary election in the manner provided for in clause 2 above for their representation in Provincial Legislatures.
4. In the Central Legislature 18 per cent of the seats allotted to the general electorate for British India in the said legislature shall be reserved for the Depressed Classes.
5. The system of primary election to panel of candidates for election to the Central and Provincial Legislatures, as herein before mentioned, shall come to an end after the first ten years unless terminated sooner by mutual agreement under the provision of Clause 6 below.

6. The system of representation of the Depressed Classes by reserved seats in the Provincial and Central Legislatures as provided for in clauses I and 4 shall continue until determined by mutual agreement between the communities concerned in this settlement.
7. The franchise for the Central and Provincial Legislature for the Depressed Classes shall be as indicated in the Lothian Committee Report.
8. There shall be no disabilities attaching to anyone on the ground of his being a member of the Depressed Classes in regard to any elections to local bodies or appointment to public service. Every endeavour shall be made to secure a fair representation of the Depressed Classes in these respects subject to such educational qualifications as may be laid down for appointment to public services.
9. In every province out of the educational grant an adequate sum shall be earmarked for providing educational facilities to members of the Depressed Classes. The terms of the Pact were accepted by Mr. Gandhi and given effect to by Government by embodying them in the Government of India Act.

This fast unto death was a great gamble on the part of Mr. Gandhi. He perhaps felt that the mere threat to fast unto death would make me and other Depressed Classes who were with me just shiver and yield. But he soon found that he was mistaken and that the Untouchables were equally determined to fight to the last for their rights. No one except his own followers was convinced that Mr. Gandhi's fast had any moral basis and if Gandhi got a second lease of life, he owes it entirely to the generosity and goodwill shown towards him by the Untouchables.

Question however is what advantage the Untouchables have got by entering into the Poona Pact. To understand this one must examine the results of the elections to the Legislatures. The Government of India Act came into operation on 1st April 1937. In February 1937 the elections to the new legislatures as defined in the Act took place. So far as the Untouchables are concerned the elections which took place in February 1937 were elections in accordance with the Poona Pact. The following is the analysis of the results of that election to the seats reserved for the Untouchables in the different Provincial Assemblies.

	Total Seats	*Total Seats*
Province	*Reserved for the Untouchables*	*Captured by the Congress*
United Provinces	20	16
Madras	30	26
Bengal	30	6
Central Provinces	20	7
Bombay	15	4
Bihar	15	11
Punjab	8	Nil
Assam	7	4
Orissa	6	4
Total	151	78

This analysis reveals certain facts which make one ask whether the Untouchables have got anything of any value by entering into the Poona Pact and saving the life of Mr. Gandhi and whether the Poona Pact has not made the Untouchables the Bondsmen of the Caste Hindus.

This analysis shows that a large majority of them have been elected as Congressmen. It is my firm conviction that for the Untouchables to merge in the Congress or for the matter of that in any large political party cannot but be fatal for them.

The Untouchables need a movement if they are to remain conscious of their wrongs and if the spirit of revolt is kept alive amongst them. They need a movement because the Caste Hindus have to be told that what is tragedy of the Untouchables is their crime. The Congress may not be a red-blooded Hindu body so far as the Musalmans are concerned. But it is certainly a full-blooded and blue-blooded Hindu body inasmuch as it consists of Caste Hindus. A movement of the Untouchables must mean an open war upon the Caste Hindus. A movement of the

Untouchables within the Congress is quite impossible. It must mean an *inter necine* within the party. The Congress for its own safety cannot allow it.

The Congress has strictly forbidden the Untouchables who have joined the Congress to carry on any independent movement of the Untouchables not approved of by the High Command. The result is that in those Provinces where the Untouchables have joined the Congress the movement of the Untouchables as such is dead.

The Untouchables must retain their right to freedom of speech and freedom of action on the floor of the Legislature if they are to ventilate their grievances and obtain redress of their wrongs by political action. But this freedom of speech and action has been lost by the representatives of the Untouchables who have joined the Congress. They cannot vote as they like, they cannot speak what they think. They cannot ask a question, they cannot move a resolution and they cannot bring in a Bill. They are completely under the control of the Congress Party Executive. They have only such freedom as the Congress Executive may choose to allow them. The result is that though the tale of woes of the Untouchables is ever-increasing, the untouchable members of the Legislature are unable even to ask a question about them.

So pitiable has their condition become that the Congress Party sometimes requires them to vote against a measure that may in the opinion of the Untouchable members of the Legislature be beneficial to the Untouchables. A recent instance of this occurred in Madras. Rao Bahadur Raja a member of the Madras Legislature brought in a Bill to secure the entry of the Untouchables into Hindu Temples in the Madras Presidency. The Congress Government had promised to support it at first. Subsequently the Congress Government in Madras changed its opinion and opposed the measure. It was a dilemma for the Untouchable members of the Madras Legislature. But they had no choice. The whip was applied and they in a body voted against the measure. The representatives of the Untouchables were supposed to be the watch-dogs of the Untouchables. But by reason of having joined the Congress they are muzzled dogs. Far from biting they are not even able to bark. This loss of freedom of speech and action by these Untouchable members is entirely due to their having joined the Congress and subjected themselves to the discipline of the Congress.

The third disadvantage arising from the Untouchables joining the Congress lies in their being unable to secure any real benefit to the Untouchables. This is due to two reasons. First of all the Congress is not a radical party. The Congress has the reputation of being a revolutionary organization. Its idea of complete independence, the movement of civil disobedience and non-payment of land revenue which the Congress once launched have undoubtedly given that reputation. But many people forget that a revolutionary party is not necessarily a radical party. Whether a revolutionary party is also a radical party must depend upon the social and emotional realities which bring on or induce the revolutionary activity.

The Barons of England who under Simon de Mandfort rose against King John in 1215 and compelled him to sign the *Magna Charta* must be classed as revolutionaries along with the Peasants of England who in 1381 rose in rebellion under Wat Tyler against their masters and who were all hanged for their rebellious acts. But who can say that the Barons because they were revolutionaries were also radical? The Barons rebelled because they wanted the rights of their class against the King and the peasants established. The Barons revolt was fed by the social emotion of those who were frustrated of power. The emotions behind the peasants revolt were those who were oppressed and who were hungering for food and freedom and that is why the peasants were both revolutionary as well as radical. The revolt of the Congress is more like the revolt of the Peasants.

The Congress under Gandhi is as radical as the Barons were under Simon de Mandfort. Just as the Barons revolt was fed by the social emotions of those who were frustrated of power and not by the emotions of those who were toiling and hungering so was the Congress revolt against the British. It is true that the Congress gathered a large following from the masses but that was by appealing to their anti-British feeling which is natural to all Indians. It is also true that emotions of these were those who were frustrated of food and freedom. But their emotions were in conflict of those socially advanced and the propertied classes. And the latter had all along been the governing class in the Congress. The masses have been camp followers. It is their emotions which has all along determined the character of the Congress. Their emotions are of those who are frustrated of power. That is why the Congress has been only revolutionary body and has not been a radical party. The truth of this can be seen by any one who cares to examine the record of the Congress Governments. Their achievement since they have taken over are just a miscellaneous collection of trifling trinkets. They have shot down the workers more readily than the British and have released criminals sentenced by the High Courts on no other ground than that they have the authority to do it. It is a surprize to me—it is not me— that the Congress has so soon shown that it is just a counter-part of the Tories in England. The governing class in the Congress has lost all its fervour for revolution, for driving the British out. Having now got a field to exploit the masses they want to stick on the power and authority to do the job thoroughly and do not wish to be disturbed by any thought of anti-imperialism at all.

Not being radical party the Congress cannot be trusted to undertake a radical programme of social and economic reconstruction without which the Untouchables can never succeed in improving their lot. For the Untouchables to join such a party is a futile and senseless thing. The Congress will not do anything for them but will only use them as they have done. The Congress might do something for the Untouchables if it was compelled to do by force of circumstances. There is only one circumstance in which the Congress would feel such a compulsion—that is when the Congress finds itself dependent upon the representatives of the Untouchables for its majority in the Legislature. Then the Untouchables would be in a position to dictate their terms to the Congress and the Congress would be bound to accept them. In such a contingency it would be worth the while of the Untouchables to join the Congress in a coalition. It would be real bargain. But today the Congress has everywhere such large majorities that in the legislatures it is its own master. It is not dependent on any outside support. The Untouchables who are in it are at the end of the tail and the tail so lengthened that it cannot wag. This is the second reason why joining the Congress can be of no benefit to the

Untouchables. Such are the disadvantages that have arisen from the Untouchables having joined the Congress. They are not merely disadvantages. I call them dire consequences. All social movement has become dead. All political power has become migratory. It is understatement to say that under the new Constitution the Untouchables are marking time. The fact is that they have been put in chains.

But the question will undoubtedly be asked and it is this—If such are the consequences of joining the Congress why did these Untouchables join it? Why did they not fight their elections independently and in opposition to the Congress? Some of the Untouchables who stood on the Congress ticket were just careerists, men on the make who wanted to climb into the Legislature so as to be within call when the places of office or profit come for distribution. They did not care by whose ladder they climbed. The Congress being the biggest party and its pass the surest way of being admitted into the Legislature these careerists felt that to join the Congress was the easiest way of electoral success. They did not want to take any chance. This however explains their object. It does not explain the cause which forced them to join the Congress. I am sure even these careerists would not have joined the Congress if it was possible for them to have got themselves elected independently of the Congress. They joined the Congress only because they found that course was impossible. Why were they compelled to join the Congress? The answer is that it was due to the system of joint electorates which caused the mischief which was introduced by the Poona Pact.

A joint electorate for a small minority and a vast majority is bound to result in a disaster to the minority. A candidate put up by the minority cannot be successful even if the whole of the minority were solidly behind him. The fact that a seat is reserved for a minority merely gives a security that the minority candidate will be declared elected. But it cannot guarantee that the minority candidate declared elected will be a person of its choice if the election is to be by a joint electorate. Even if a seat is reserved for a minority, a majority can always pick up a person belonging to the minority and put him up as a candidate for the reserved seat as against a candidate put up by the minority and get him elected by helping its nominee with the superfluous voting strength which is at its command. The result is that the representative of the minority elected to the reserved seat instead of being a champion of the minority is really the slave of the majority.

In the system of electorates now formed for election to the Legislature, the Untouchable voters as against the caste Hindu voters are placed in a hopeless minority. A few instances will show how great is the discrepancy in the relative voting strength of the Untouchables and the caste Hindus in the different constituencies.

The power to do mischief in elections which a joint electorate gives to a majority is increased immensely if the electoral system is based on the principle of a single member constituency.

In a system of joint electorates with reserved seats for a minority a constituency must always be a plural member constituency *i.e.* there must be one seat for the minority and at least one seat for the majority. In other words it must be what is called a plural member constituency. This plural member constituency must be small one *i.e.* the majority community may have just two seats as against the one assigned to the minority. It may be a large one *i.e.* the majority may have a larger number of seats assigned to it. This is an important consideration because the smaller the number of seats the greater the power of mischief which the majority gets.

This will be clear if it is borne in mind that when a majority has fewer seats it can release a large portion of its voting strength to get its own nominee from the minority elected to the reserved seats and defeat the nominee of the minority. On the other hand if the majority is assigned a larger number of seats, there a competition among the candidates is greater, the voters of the majority community are for the most part (busy) in fighting out the election to the seats assigned to the majority and very few at all can be released to help the nominee of the majority for the minority seat. In a joint electorate the safety of the minority lies in the majority having a larger number of seats to contest. Otherwise it is sure to be overwhelmed by the majority.

In the electoral system now framed for the caste Hindus the principle that is adopted is that of the single member constituency. It is true that on the face of it the constituency taken as a whole appear to be a plural member constituency. But, in fact, the constituency so far as the caste Hindus are concerned are single member constituencies. The consequence of this single member constituency system for Caste Hindus is that the Hindus are able to release an enormous lot of their superfluous votes and flood the election for the seats reserved for the Untouchables and keep their nominee for the reserved seat afloat.

The Hindus were anxious to forge further means for nullifying the benefit of the Poona Pact. The Poona Pact having been concluded in a hurry, it left many things undefined. Of the things that were left undefined and about which there arose subsequently a keen controversy were the following: (1) Does the 'panel of four' to be elected at the primary election imply four as a maximum or a minimum? (2) What was to be the method of voting in the final joint electorate with the Hindus? The Hammond Committee which had to decide upon these issues found that there were two diametrically opposite views in regard to these two questions, one view held by the Caste Hindus and the other held by the Untouchables.

It was contended on behalf of the Caste Hindus that the panel of four was intended to be a minimum. If four candidates are not forthcoming there could be no primary election and therefore there can be no election for the reserved seat, which they said must remain vacant and the Untouchables should go without representation. The Untouchables contended that four was the maximum. Four in the Poona Pact meant "not more than four". It did not mean "not less than four". On the question of voting the caste Hindus contended that the compulsory distributive vote was the most appropriate. The Untouchables on the other hand insisted that the cumulative system of voting was the proper system to be introduced.

The Hammond Committee accepted the view propounded by the Untouchables and rejected those of the Caste Hindus. All the same it is interesting to know why the caste Hindus put forth their contentions.

The reason why the Hindus wanted four in the panel and not less was quite obvious. The object of the Hindus is to get elected in the final election such a representative of the Untouchables as would be most ready and willing to compromise with Hindus and Hinduism. To get him elected in the final election he must first come in the Panel. A most compromising Untouchable can come in the Panel only when the panel is a large panel.

If there is only one candidate in the Panel then he would be the stautichest representative of the Untouchable and worst from the standpoint of the Hindus. If there are two, the second will be less staunch than the first and therefore good from the standpoint of the Hindus. If there are three, the third will be less staunch than the second and therefore better from the standpoint of the Hindus. If there be four, the fourth will be less staunch than the third and therefore best from the point of view of the Hindus. The Panel of four gives the Hindus the best chance of getting into the Panel such representative of the Untouchables as is most compromising in his attitude towards Hindus and Hinduism and that is why they insisted that the Panel should be at least of four.

The object of insisting upon the system of compulsory distributive vote was just supplementary to the idea of having the Panel of not less than four. Under the cumulative vote the elector has as many votes as there are seats, but may plump them all for one candidate or distribute them over two or more candidates as he may desire. Under the distributive system of voting the elector has also as many votes as there are seats, but he can give only one vote to any one candidate.

Although the two look different yet in effect there may be no difference because even under the cumulative vote a voter is not prevented from distributing his votes. He is free to give one vote to one candidate. But the Hindus did not want to take any chance. Their main object was to flood the election to the seat reserved for the Untouchables in the joint electorate by using the surplus votes of the caste Hindus in favour of the Untouchable candidate who happens to be their nominee. The object was to outnumber the Untouchable voters and prevent them from electing their own nominee. This cannot be done unless the surplus votes of the caste Hindu voters were divested from the caste Hindu candidate towards the Untouchable candidates. There is a greater chance of the diversion of these surplus votes under the distributive system than there is under the cumulative system. Under the former the caste Hindu voter can give only one vote to the caste Hindu candidate.

The other vote not being of use to the caste Hindu candidate is usable only for an Untouchable candidate. With the distributive system there was more chance of flooding the election to the seat reserved for the Untouchables and this is why they preferred it to the system of cumulative vote. But they want to take a chance. Even the distributive system from their point of view was not foolproof. Under the distributive system there was no compulsion upon the voter to use all his votes. He may use one vote for the caste Hindu candidate and not use the rest of his votes. If this happens the purpose of getting in their untouchable nominee would be defeated. Not to leave things to chance the Hindus wanted that the distributive system of voting should be made compulsory so that a caste Hindu voter whether he wants it or not can have no option but to vote for the untouchable candidate who may be the nominee of the Hindus. The two proposals were thus a part of a deep conspiracy on the part of the Hindus. They were rejected by the Hammond Committee. But there are enough elements of mischief in the Poona Pact itself that the rejection of these two proposals has in no way weakened the power of the Hindus to render nugatory the right of special representation granted to the Untouchables.

Notwithstanding the political disaster which has overtaken the Untouchables and which is the result of the Poona Pact, there are not wanting friends of Mr. Gandhi who hold out the Poona Pact as a great boon to the Untouchables.

Firstly it is alleged that the Poona Pact gave the Untouchables larger number of seats than was given to them by the Communal Award. It is true that the Poona Pact gave the Untouchables 151 seats while the Award had only given them 78. But to conclude from this that the Poona Pact gave them more than what was given by the Award is to ignore what the Award had in fact given to the Untouchables.

The Communal Award gave the Untouchables two benefits: *(i)* a fixed quota of seats to be elected by separate electorates for Untouchables and to be filled by persons belonging to the

Untouchables, (ii) Double vote, one to be used through separate electorates and the other to be used in the general electorates.

Now if the Poona Pact increased the fixed quota of seats it also took away the right to the double vote. This increase in seats can never be deemed to be a compensation for the loss of the double vote. The second vote given by the Communal Award was a priceless privilege. Its value as a political weapon was beyond reckoning. The voting strength of the Untouchables in each constituency is one to ten. With this voting strength free to be used in the election of caste Hindu candidates, the Untouchables would have been in a position to determine, if not to dictate, the issue of the General Election. No Caste Hindu candidate could have dared to neglect the Untouchables in his constituency or be hostile to their interest if he was made dependent upon the votes of the Untouchables. Today the Untouchables have a few more seats than were given to them by the Communal Award. But this is all that they have. Every other member is indifferent if not hostile. If the Communal Award with its system of double voting had remained, the Untouchables would have had a few seats less but every other member would have been a member for the Untouchables. The increase in the number of seats for the Untouchables is no increase at all.

Admitting for the sake of argument that the Poona Pact did give to the Untouchables a few more seats, the question still remains of what use are these additional seats. Ordinarily a right to vote is deemed to be sufficient as a means of political protection. But it was felt that in the case of the Untouchables mere right to vote would not be enough. It was feared that a member elected on the votes of the Untouchables, if he is himself not an Untouchable, might play false and might take no interest in them. It was held that the grievances of the Untouchables must be ventilated in the Legislature and that the surest way of ensuring it would be to provide that a certain number of seats shall be reserved for them so that the Untouchables shall be represented by Untouchables in the Legislature. It is now evident that this hope has not been fulfilled. The Communal Award no doubt gave fewer seats. But they would have been all of them freemen. The Poona Pact gave more but they are all filled by bondsmen. If to have a platoon of bondsmen is an advantage then the Poona Pact may be said to be advantage.

The second argument in favour of the Poona Pact is that in abolishing separate electorates it saved the Untouchables from being eternally branded as Untouchables. In one of his speeches delivered in London Mr. Gandhi said:

"Muslims and Sikhs are all well organized. The 'untouchables' are not. There is very little political consciousness among them and they are so horribly treated that I want to save them against themselves. If they had separate electorates their lives would be miserable in villages which are the strongholds of Hindu orthodoxy. It is the superior class of Hindus who have to do penance for having neglected the ' untouchables ' for ages. That penance can be done by active social reform and by making the lot of the ' untouchables ' more bearable by acts of service, but not by asking for separate electorates for them. By giving them separate electorates you will throw the apple of discord between the 'untouchables' and the orthodox. You must understand I can tolerate the proposal for special representation of the Musalmans and the Sikhs only as a necessary evil. It would be a positive danger for the ' untouchables '. I am certain that the question of separate electorates for the ' untouchables ' is modern manufacture of Government. The only thing needed is to put them on the voters' list, and provide for fundamental rights for them in the constitution. In case they are unjustly treated and their representative is deliberately excluded, they would have the right to special election tribunal which would give them complete protection. It should be open to these tribunals to order the unseating of an elected candidate and the election of the excluded man."

"Separate electorates to the 'untouchables' will ensure them bondage in perpetuity. The Musalmans will never cease to be Musalmans by having separate electorates. Do you want the 'untouchables' to remain 'untouchables' for ever? Well, the separate electorates would perpetuate the stigma. What is needed is destruction of untouchability, and when you have done it, the bar

sinister which has been imposed by an insolent ' superior ' class upon an ' inferior ' class will be destroyed. When you have destroyed the bar sinister, to whom will you give the separate electorates? Look at the history of Europe. Have you got separate electorates for the working classes or women? With adult franchise, you give the ' untouchables ' complete security. Even the orthodox would have to approach them for votes."

"How then, you ask, does Dr. Ambedkar, their representative, insist on separate electorates for them? I have the highest regard for Dr. Ambedkar. He has every right to be bitter. That he does not break our heads is an act of self-restraint on his part. He is today so much saturated with suspicion that he cannot see anything else. He sees in every Hindu a determined opponent of the ' untouchables ', and it is quite natural. The same thing happened to me in my early days in South Africa, where I was hounded out by the Europeans wherever I went. It is quite natural for him to vent his wrath. But the separate electorates that he seeks will not give him social reform. He may himself mount to power and position but nothing good will accrue to the ' untouchables '. I can say all this with authority, having lived with the 'untouchables' and having shared their joys and sorrows all these years."

His argument it is true derives some support from the Simon Commission which also observed: (*The following extracts are not quoted in the MS. They are reproduced here from Dr. Ambedkar's 'What Congress and Gandhi have done to the Untouchables', Appendix VI, p. 327.*— Ed.)

Extracts from the Report of the Simon Commission:

78. In no other province has it been possible to get an estimate of the number of the depressed classes who are qualified to vote. It is clear that even with a considerable lowering of the franchise—which would no doubt increase the proportion of the depressed class voters—there would be no hope of the depressed classes getting their own representatives elected in general constituencies without special provision being made to secure it. In the long run the progress of the depressed

classes, so far as it can be secured by the exercise by them of political influence, will depend on their getting a position of sufficient importance for other elements to seek their support and to consider their needs.

80. It will be seen, therefore, that we do not recommend allocating seats to the depressed classes on the basis of their full population ratio. The scale of reserved representation suggested will secure a substantial increase in the number of the M.L.C.s drawn from the depressed classes. The poverty and want of education which so widely prevail amongst them make it extremely doubtful whether a large number of adequately equipped members could be at once provided, and it is far better that they should be represented by qualified spokesmen rather than by a larger number of ineffectives who are only too likely to be subservient to higher castes. The re-distribution of seats which is now being attempted among different kinds of representatives cannot be permanent, and provision must be made for its revision. But we think that our proposal is adequate for the present, especially as the representation of opinion by reservation of seats does not exclude the possibility of the capture of other seats not so reserved.

But that this argument is silly there can be no doubt about it. To put a man in a separate category from others is not necessarily an evil. Whether the affixing of a label is good or bad depends upon the underlying purpose. If the object is to deprive him of rights then such a labelling would no doubt be a grievous wrong. But if the purpose is to mark off as a recipient of a privilege then far from being a wrong it would be a most beneficial measure. To enrol an untouchable in a separate electoral roll would be objectionable if the object was to deprive him of the right of franchise. To enrol him in a separate electoral roll for giving him the benefit of special representation would certainly be an advantage to him. Looked at from the point of view of ultimate purpose it is difficult to see how any person who claims to be the friend of the untouchables could object to separate electorates for them. Not only the argument of Gandhi against separate electorate was silly it was also insincere.

Gandhi objected to separate electorates because it involved labelling of the Untouchables. But how is this labelling avoided injoint electorates it is difficult to understand. The reservation of a seat for the Untouchables in a joint electorate must and does involve such labelling for the candidate claiming the benefit of the reserved must in law declare that he is an untouchable. To that extent there is certainly a labelling involving in the Joint Electorate and Mr. Gandhi should have objected to joint electorates as violently as he did to separate electorates., Either Mr. Gandhi was insincere or that he did not know what he was talking about.

Friends of Mr. Gandhi do not stop to consider how far under the Poona Pact the Untouchables have been able to send independent men to represent them in the Legislature and whether these representatives have been putting up any fight and how well they are succeeding. If they stopped to do it they would be ashamed to sing the praises of the joint electorates and the Poona Pact. The Congress and the Hindus have shamefully abused their power and their resources as a majority community. Not only have they prevented the Untouchables from electing persons of their choice, not only have they got their own nominees elected by the use of their surplus votes, they have done something for which any decent party in any part of the world ashamed of itself. The selection of the candidates from the untouchables made by the Congress to fill the seats reserved for the Untouchables was a most cowardly and a blackguardly act.

It was open to the Congress—which is simply a political surname for Hindus—to allow the Untouchables the benefits of more seats than those fixed for them by the Poona Pact. They could have done that by adopting untouchable candidates to contest general seats. There was nothing in law to prevent them from doing so. The Congress did nothing of the kind. This shows that if no seat had been reserved to the Untouchables, the Hindus would never have cared to see that an untouchable was returned to the legislature. On the other hand when seats were reserved, the Hindus came forward to spoil the effect of the reservation by seeing to it that the seats went to such untouchables as agreed to be their slaves.

Thus there has been a tragic end to this fight of the untouchables for political rights. I have no hesitation in saying that Mr. Gandhi is solely responsible for this tragedy.

Mr. Gandhi's cry against the Communal Award on the ground that it prescribed separate electorates was absolutely false and if the Hindus had not become maddened by his fast they would have seen that it was so. The Communal Award had also a provision for joint electorates in addition to separate electorates. The second vote given to the Untouchables was to be exercised in a general electorate in the election of a Caste Hindu candidate. This was undoubtedly a system of joint electorates. The difference between the Communal Award and the Poona Pact does lie in the nature of electorates. Both provide for joint electorates. The difference lies in this, joint electorate of the Communal Award was intended to enable the Untouchables to take part and influence the election of the Caste Hindu candidate while the joint electorate of the Poona Pact is intended to enable the caste.

Hindus to take part and influence the election of the Untouchable candidates. This is the real difference between the two.

What would a friend of the Untouchables wish? Would he support the joint electorates of the Communal Award or the joint electorates of the Poona Pact? There can be no doubt that the real salvation of the Untouchables lies in making the Hindus dependent upon the suffrages of the Untouchables. This is what the Communal Award did. To make the Untouchables depend upon the suffrages of the Hindus is to make them the slaves of the Hindus which they already are. This is what the Poona Pact does. The Communal Award was intended to free the Untouchables from the thraldom of the Hindus. The Poona Pact is designed to place them under the domination of the Hindus.

This 'Fast unto Death' of Mr. Gandhi was described in glorious terms by his friends and admirers both in India and outside. It was described as ' second crucifixion ', as ' martyrdom ' and as ' Triumphal struggle'. An American friend of Mr. Gandhi assured the Americans that in laying down his life Mr. Gandhi was neither a ' trickster ' nor a stick demagogue. Another American in his ecstasy went to the length of describing him as the incarnation of *'one against the world'*. Of course I was held out as the villain of the piece. I had of course my own view of Gandhi's fast. I described it as a political stunt. His utterances had to me always the ring of falsity and even of insincerity.

I had always the feeling that what actuated Mr. Gandhi to fast against the Communal Award was not any desire to liberate the Untouchables as to save the Hindus from disruption. He was prepared to do that at any cost, even at the cost of political enslavement of the Untouchables. His disapproval of the Poona Pact was very much like the disapproval of the enfranchisement of the Negro by the Southerners after the civil war. The ' Statesman ' and 'Nation ' came to the same conclusion.

At the time there was this one solitary instance of a view agreeing with mine. Even some of the prominent untouchables backed Mr. Gandhi. A curious case was that of Mr. Raja whose grievance was that although he was a member of the Central Assembly nominated to represent the Depressed Classes he was not selected as a delegate for the Round Table Conference. He was fighting for separate electorates. Suddenly he changed sides and took up the cudgels on behalf of Mr. Gandhi and fulminated both against me for demanding and against the British Government for granting separate electorates. He developed such a strong love for Mr. Gandhi and such a strong faith in the Hindus that no one could suspect that he was doing the work of a mere hireling. This is what Mr. Raja said in the course of a speech delivered by him on an adjournment motion moved in the Central Legislature on September 13th, 1932 relating to Gandhi's fast.

> *"Never in the annals of the history of India has the issue of the Depressed Classes assumed importance as it has today, and for this we of the Depressed Classes must for ever be grateful to Mahatma Gandhi. He has told the world, in words which cannot be mistaken, that our regeneration is the fundamental aim of his life. If world conscience cannot be roused even now to the realization of the position of the Depressed Classes, then we can only conclude that all instincts of humanity are dead in the world today."*
>
> *"The question before the House is the situation created by Mahatma Gandhi opposing the grant of communal electorates to the Depressed Classes. I am sure there is no honourable member in this House who will not regret*

that circumstances should have forced such a great personality to take a vow to play on his life, but sir, the correspondence shows that Government had enough warning. If they did not attach full importance to our considered views expressed in our conferences and in the Rajah-Moonje Pact I had signed with the President of Hindu Mahasabha, they should have taken at least the grave warning given by Mahatma Gandhi and desisted from the course of creating separate electorates."

"Indeed this is my chief attack on the Premier's letter to Mahatma Gandhi. He tells us that he has given separate electorates for twenty years to enable us to get the minimum number of seats to place our views before the Government and legislature of the day. I contend that this privilege we have already enjoyed under the Montford reforms which have enabled us to get representation in numerous local bodies and in legislatures both provincial and central. We are sufficiently organized for that purpose and do not need either special pleading or special succour. In future what we do need as real remedy for our uplift is definite power to elect our representatives from the general constituencies and hold them responsible to us for their actions. I do not know why the Prime Minister calls the scheme of joint electorates with reservation of seats as impracticable. It is already in force in local bodies in Madras and some other provinces and has worked very satisfactorily. I contend, sir, that the scheme enunciated in the communal decision involves our segregation and makes us politically untouchables. I am surprised at the argument of the Prime Minister that there is no segregation because we can vote for Caste Hindus who will have to solicit our votes. But, sir, how can we bring about common ideal of citizenship when Depressed Class representatives are not to solicit votes of higher castes?"

"The sufferings which my community has undergone at the hands of Caste Hindus have been acknowledged by Caste Hindus themselves, and I am prepared to admit that there are a large number of reformers among them who are doing everything possible to improve our status

and position. I am convinced that there is a change of heart and a change in the angle of vision of Caste Hindus. We, Depressed Classes, feel ourselves as true Hindus as any Caste Hindu can be, and we feel that the moral conscience of the Hindus has been roused to the extent that our salvation lies in bringing about a change from within the main body of Hindu society and not segregating ourselves from them. The course adopted by the Government would certainly arrest the progress of this most laudable movement. I must say, sir, that the Prime Minister's letter in its entire conception, and expression has disappointed me."

"The crisis that faces us today is very grave. There hangs in the balance the life of the greatest Indian of our time, and there hangs in the balance the future of millions of the down-trodden people of this country. Is Government going to take the responsibility for killing the one and reducing the other to perpetual servitude? Let it make its choice well and wisely."

Mr. Raja not only backed the Poona Pact and fought for distributive vote which as I pointed out was nothing but a part of the design of the Hindus to make the political enslavement of the Untouchables fool-proof and knave-proof. Mr. Raja was so much enamoured of his new faith in Gandhi and Hindus that he was not satisfied by the disposal of the matter by the Hammond Committee. He reopened the matter after election by moving a resolution in the Madras Assembly in favour of the distributive vote1.

But today after seeing the results of the Poona Pact Mr. Raja seems to have been disillusioned. How long he will remain faithful to the truth he has discovered is more than I can say. But he has declared himself openly as a bitter opponent of the Poona Pact. In a letter to Mr. Gandhi, dated 25th August 1938, Mr. Raja says:

"You remember how, when most of my people were in favour of separate electorates so that they may express themselves faithfully and effectively in the legislatures, you staked your life on bringing them into the Hindu fold not only politically but socially and religiously. And I was in no small measure responsible for my

people going in for a joint electorate with Caste Hindus on the express understanding that there was to be no interference with our choosing men who would faithfully represent our grievances and wishes. It was with this object that the panel election was instituted."

"All this you know as well as I do. But my object in recalling the fact is to show that while on our part we faithfully adhered to the Poona Pact, giving up agitating for a separate electorate, the Congress Party men in Madras representing the Caste Hindus deviated from the Pact, so much so, that our community in the Legislative Assembly have to follow the Caste Hindus blindly in Madras Legislative Assembly Debates. Every measure the Ministry may bring and vote with them even in matters which deeply affect the interests of the community."

"You perhaps remember that at the beginning of the elections I protested against the Congress Committee setting up candidates for the panel election among the Depressed Classes. You were good enough to say that I might allow my community joining the Congress on certain conditions placed before us by Mr. S. Satyamurti. One of these conditions was that in matters affecting their community, the Depressed Class members of the Congress Party need not vote with the Congress members but vote according to their judgement."

"The recent debate on the Temple Entry Bill in the Madras Legislative Assembly has exposed the ugly fact that all Depressed Class Members driven by the discipline of the Congress Party in the Assembly voted solidly against my motion for referring the Bill to a Select Committee. Could anything be more unnatural and more humiliating, proving as it did the subjugation of my community by the Caste Hindus represented by Mr. Rajagopalachariar?"

"You know the provisions of the Bill. It was only a piece of permissive legislation making it possible for a majority of worshippers of a temple to allow Harijans to worship in the temple. There was no element of compulsion or coercion in it. This Bill had your blessing. It was drafted

by Mr. Rajagopalachariar himself and approved by you."

"At a previous session of the Assembly I introduced the Bill with the consent of Mr. Rajagopalachariar, who promised his full support to the measure. When I suggested that the Bill might be introduced by him as a Government measure, he wanted me to introduce it. When I met him last, on the 12th July 1938 and informed him that I was giving notice of a motion for referring the Bill to a Select Committee he did not object."

"I do not know what happened in the meantime but two days before my motion for referring the Bill to the Select Committee came up before the house, Mr. Rajagopalachariar sent for me and quietly asked me to withdraw the Bill, which I refused to do. When in due course, I moved for the consideration of the Bill, Mr. Rajagopalachariar stood up and opposed the Bill and requested me to withdraw it, saying that he would introduce a Temple Entry Bill on the same lines, only for Malabar and not for other Districts."

"The effect of Mr. Rajagopalachariar's speech was to defeat my motion with my own community men registering their votes against the measure, introduced to secure their social and religious elevation. One effect of Mr. Rajagopalachariar's opposition will be to strengthen the opposition in the country to temple entry as a whole."

"All this makes me uneasy about the wisdom of our havirig been parties to the Poona Pact in the full belief that the Congress would really help in our attempt to secure social and religious freedom. I am forced to think that our entering the joint electorate with the caste Hindus under the leadership of the Congress, far from helping us, has enabled the Congress, led by Caste Hindu leaders to destroy our independence and to use us to cut our own throats."

"In the course of the debate, I asked Mr. Rajagopalachariar whether he had obtained your approval of the attitude which he so suddenly and unexpectedly assumed and the Speaker said that Mr.

Rajagopalachariar will be given an opportunity to answer the question after I had done. Though he was given the privilege of speaking after me, he avoided the question and did not answer it at all."

"I trust that you will give your most serious consideration to this question of the attitude of the Congress Ministry in Madras towards Harijan uplift, especially with regard to temple entry and let me have your view before I answer my community men who are very much exercised over this question and are contemplating a repudiation of the Poona Pact and an agitation for a separate electorate accompanied by direct action of some kind." Mr. Rajah also wired to Mahatma Gandhi, Wardha on 12th September 1938 as follows:

"Agitation against Ministry rejection my bill growing difficult withstand pressure upon me—Anxiously awaiting reply."

Mr. Gandhi replied to Mr. Rajah on 14th September 1938: "Dear friend, I must apologise for the delay in replying to your letter, I have been overwhelmed with work. Now I have your wire.

I wish you would trust C. Rajagopalachariar to do his best. He should be allowed to do the thing in his own way. If you cannot trust, naturally you will take the course which commends itself to you. All I know is that Harijans have no better friend than him. Go to him, reason with him and if you cannot persuade him, bear with him. That is my advice."

Mr. Rajah wrote a letter dated 21st September 1938 to Mr. Gandhi stating: "I should request you on the other hand to give more serious consideration to the pledges given to my community during the Yeravada Fast and to the way in which they are carried out by your plenipotentiary in Madras. That fast was undertaken by you in order to change the Communal Award providing separate electorates for my people and to bring them into joint electorate with Caste Hindus by promising to spare no efforts to remove untouchability. And you have more than once said that temple entry is of the very essence of the removal of untouchability. So the question of our being in a joint electorate with Caste Hindus and the attitude of the Congress Ministry towards the raising of the social and religious status of the community are mutually and vitally connected.

If we are not free to enter into Hindu Temples, we are no Hindus, and if we are not Hindus why should we be in a joint electorate with them? Is it for swelling their numbers as against Muslims and other communities?

If you look at the situation in Madras from this point of view, you will realise that the rejection of the Temple Entry Bill is a gross betrayal of the Depressed Classes by the Congress Government in Madras.

Any amount of money spent out of public funds or even from private resources for the amelioration of the economic condition of my community will be no substitute for the removal of untouchability through temple entry. As you yourself once said temple entry is the acid test of the sincerity of the caste Hindus in espousing the cause of the Depressed Classes.

Mr. Rajagopalachariar's Temple Entry Bill, besides involving the tactical choice of a difficult district makes our community subservient to the will of Caste Hindus, a policy which is given a further effect to, in the appointment of Advisory Boards to assist the Labour Commissioner in which Caste Hindus of the Harijan Sevak Sanghs in this Province are given Government facilities for guiding the destinies of my people. I do not expect you to agree with my views on the measures recently introduced by Mr. Rajagopalachariar after rejecting the Madras Temple Entry Bill; but I expect you to give more serious consideration to the situation in Madras as regards the Depressed Classes in the light of the pledges given to the Depressed Classes during your memorable fast.

You are morally bound to make this a matter of conscience and not merely one of political strategy. I assure that if you should seek ' inner light ' on this subject of untouchability and temple entry you will speak out more plainly and make the necessary sacrifice to educate your followers.

I propose to send this correspondence to the press, but I shall wait for any further word from you till the end of this month." Mr. Gandhi replied to the above on 5th October 1938:

I am working under great difficulty. Even this I am writing in the train taking me to Peshawar.

Of course you will publish the correspondence between us whenever you think it necessary. Your last letter shows that you are in the wrong. I am not partial to Rajaji. But I know that he is as firm on untouchability as I am myself. I must, therefore, trust his judgement as to how to do the thing. From this distance, I can't judge his action. Do you not see that the whole of the movement is one of conversion of the Sanatani heart? You cannot force the pace except to the extent that reforms immolate themselves. This process is going on vigorously.

This temple entry question is a mighty-religious reform. I would like you to apply your religious mind to it. If you will, you will give your whole hearted support to Rajaji and make his move a thorough success.

The Untouchables of U.P. have also expressed their hostility to the Poona Pact. In a Memorandum submitted to Col. Muirhead the Under Secretary of State in India they said.

All over India the Untouchables have realized that the Poona Pact has been a trap and the change of the British Government's Communal Award by Gandhi's Poona Pact is a change which in reality a change from freedom to bondage.

The Poona Pact was signed on the 24th September 1932. On the 25th September 1932 a public meeting of the Hindus was held in Bombay to accord to it their support. At that meeting the following resolution was passed: This Conference confirms the Poona agreement arrived at between the leaders of the Caste Hindus and Depressed Classes on September 24, 1932, and trusts that the British Government will withdraw its decision creating separate electorates within the Hindu community and accept the agreement in full. The Conference urges that immediate action be taken by Government so as to enable Mahatma Gandhi to break his fast within the terms of his vow and before it becomes too late. The Conference appeals to the leaders of the communities concerned to realize the implications of the agreement and of this resolution and to make earnest endeavour to fulfil them.

This Conference resolves that henceforth, amongst Hindus, no one shall be regarded as an untouchable by reason of his birth, and that those who have been so regarded hitherto will have the same right as other Hindus in regard to the use of public wells, public schools, public roads, and all other public institutions. This right shall have statutory recognition at the first opportunity and shall be one of the earliest Acts of the Swaraj Parliament, if it shall not have received such recognition before that time.

It is further agreed that it shall be the duty of all Hindu leaders to secure, by every legitimate and peaceful means, an early removal of all social disabilities now imposed by custom upon the so-called untouchable classes, including the bar in respect of admission to temples.

Mr. Gandhi felt that an organization which will devote itself exclusively to the problem of the Untouchables was necessary. Accordingly there was established on 28th September 1932 the All-India Anti-Untouchability League. The name, Gandhi thought, did not smell well. Therefore in December 1932 it was given a new name—The Servants of the Untouchables Society. That name again was not as sweet as Mr. Gandhi wished it to be. He changed and called it the Harjan Sevak Sangh.

The first change which Mr. Gandhi has brought about is this change in the name. Instead of being called Untouchables they are now called *Harijans*. To call, the Untouchables say that Mr. Gandhi is selfish and has given the name Harijan to the Untouchables to blaster up Vaishnavism. They want the Untouchables to be called Harjans the followers of Shiva. Mr. Gandhi replies that the term is used to mean God and not Vishnu and that Harijan simply means 'children of God'.

The Untouchables simply detest the name Harijan. Various grounds of objection are urged against the name. In the first place it has not bettered their position. It has not elevated them in the eyes of the Hindus. The new name has become completely identified with the subject matter of the old. Every body knows that Harijans are simply no other than the old Untouchables. The new name provides no escape to the Untouchables from the curse of untouchability. With the new name they are damned as much as they were with the old. Secondly the Untouchables say that they prefer to be called Untouchables. They argue that it is better that the wrong should be called by its known name.

It is better for the patient to know what he is suffering from. It is better for the wrong doer that the wrong is there still to be redressed. Any concealment will give a false sense of both as to existing facts. The new name in so far as it is a concealment is fraud upon the Untouchables and a false absolution to the Hindus. Thirdly there is also the feeling that the name Harijan is indicative of pity. If the name meant ' chosen people of God ' as the Jews claimed themselves to be it would have been a different matter. But to call them 'children of God' is to invite pity from their tyrants by pointing out their helplessness and their dependent condition. The more manly among the Untouchables resent the degrading implications of this new name. How great is the resentment of the Untouchables against this new name can be seen from the fact that whole body of the representatives of the Untouchables in the Bombay Legislative Assembly walked out of the House in protest when the Congress Government introduced a measure giving to the name Harijan the sanction of law.

This new name Harijan will remain until the downfall of Mr. Gandhi and the overthrow of the Congress Governments which are his creatures. That it was forced upon the Untouchables and that it has done no good are however facts which cannot be disputed.

Having discussed the blessings of this new name I must now proceed to examine the work of the Harijan Sevak Sangh. The Sangh is spoken of as a memorial to Gandhi's labours in the cause of the Untouchables. What are the achievements of the Sangh.

The Sangh is an All-India Organization. It has a Central Board. Then there are Provincial Boards and under the Provincial Boards there are District and Local Committees. The number of Provincial Boards and Local Committees is given below:

	1932-33	*1933-34*	*1934-35*
Provincial Boards	26	26	29
District & Local Committees	213	313	372

The financial resources of the Sangh are mainly drawn from the collections made on an All-India tour specially undertaken for the purpose by Mr. Gandhi between November 1933 to July 1934. The total amount collected on this tour came to about 8 lacs of Rupees and is known as the Gandhi Purse Fund and forms the principal reserve for the Sangh to draw upon. The rest of the resources are made up of annual donations.

The total expenditure of the Sangh under all heads from year to year is as follows:

1932-33	1933-34	1934-35	1935-36
Rs. 2,31,039.00	3,31,791.00	4,48,422.00	3,99354.00

In his report for 1934-35 the General Secretary reported that almost the whole of the Gandhi Purse Fund, which stood at over 8 lacs in July 1934, will be spent away by the end of the current year, *i.e.* by the end of September 1936." In the fifth Annual Report the Secretary says, "As compared to the expenditure of the previous year, there has been a reduction of more than a lac in the total expenditure. This was partly owing to the gradual exhaustion of the Gandhi Purse Fund, dislocation in the realization of local collections owing to general elections and other contributory causes. The finances of the Central office were far from satisfactory. The total expenditure of the Central office (including grants made to branches) amounted to Rs. 86,610-14-8, as against an income of Rs. 42,485-4-9 thus leaving a deficit of Rs. 44,125-9-11 which was met from the general fund. Donations for general fund amounted to only Rs. 26,173-4."

It is obvious that the Harijan Sevak Sangh is a small affair and but for the running advertisement it gets from the Press it would not even be heared off. India is a vast continent with something like 6,96,831 villages. The Untouchables are spread out all throughout these 6,96,831 villages. There is no village without its untouchables. How many Untouchables can be reached by 372 Committees. It is a tiny peck in a vast ocean. Not only its capacity to cope with the problem is limited but its resources are too meagre to permit any relief being granted to the Untouchables on an adequate scale. The Sangh has now no permanent fund.

What it had it has spent. It has to depend upon annual subscriptions. That source is also drying up leaving the Sangh with heavy deficit. The Sangh is in an exhausted condition. Its affairs in fact would have been wound up on account of its bankruptcy. If the Sangh is still existing, it is not because its endeavours are sustained by Hindu charity directed to the uplift of the Untouchables. It exists because the Congress Governments which are now established in the different Provinces have come to the rescue of the Sangh. They have handed over to the Sangh certain social welfare work which former Governments carried out through Government Departments or Government Officials with the money grants. Thus the Sangh is now living on Government funds. As an institution maintained by the Hindus with the help of Hindu Charity for the Untouchables the Sangh simply does not exist.

The din and noise which was created by Mr. Gandhi's fast was simply deafening. The readiness to make sacrifices to save his life was great and the eagerness shown to befriend the Untouchables was surprising and overwhelming. All this has vanished leaving the Untouchables high and dry. If the desire to contribute towards the maintenance of the Sangh—which the Hindus founded as an earnest of their acceptance of their obligation towards the Untouchables—is any measure of the reality of the change of heart then it must be admitted that the change has died with the occasion which caused it. Gandhi broke his fast and the Hindu lost his new-born love for the Utitouchables.

The premature decay of the Sangh should make it unnecessary for me to consider the work it did. But the Sangh is held out as a great monument to Mr. Gandhi. It is therefore proper that I should examine the work done by the Sangh and the policy underlying that work. The work of the Sangh follows certain well defined lines. In the field of education the Sangh has sought to encourage higher education among the Untouchables by instituting scholarships for the Arts, technical and professional courses. The Sangh also gives scholarships to High School students. The Sangh also maintained Hostels for Untouchable students attending colleges and High Schools. The great part of the educational activities of the Sangh is taken up in maintaining separate schools for primary stage children where there were no common schools in the vicinity or where common schools were closed to them.

Next comes the welfare activities of the Sangh. The medical aid which the Sangh undertakes to render to the Untouchables falls under this head. This is done by intenerant workers of the Sangh who go in Harijan quarters to give medical aid to the sick and ailing among the Untouchables. The Sangh also maintains a few dispensaries for the use of the untouchables. This is a very small activity of the Sangh. The more important part of the welfare activity of the Sangh relates to water supply. The Sangh does this by (1) sinking new wells or installing tube wells and pumps for the use of the untouchables, (2) repairing old ones and (3) persuading local Governments and bodies to sink and repair wells for the Untouchables.

The third line of activity undertaken by the Sangh is economic. The Sangh seems to run a few industrial schools and it is claimed that the industrial schools run by the Sangh produced a number of trained artizans who have taken to independent living. But according to the report, more successful and substantial work was done by way of organising and supervising cooperative societies among the Untouchables.

Such is in brief the record of the work done by the Sangh. It is largely directed by the Caste Hindus. There are very few Untouchables who have any voice in directing the activities of the Sangh. I have had no connection with the Sangh. But I might mention that when the Sangh was started I was invited to join. I had great desire to cooperate with the Hindus for the removal of untouchability. I had my own views regarding the policy and programme which the Sangh should adopt for accomplishing this task. Immediately after the Sangh was established I had to go to London to attend the Round Table Conference and had no opportunity to talk the matter over with the other members of the Sangh. But I posted a letter to the General Secretary of the Sangh Mr. Thakkar on the 14th November 1932 on Board the Ship *M. V. "Victoria"*. Excepting a short introductory para which I omit, the following is the full text of the letter:

In my opinion there can be two distinct methods of approaching the task of uplifting the Depressed Classes. There is a school which proceeds on the assumption that the fate of the individual belonging to the Depressed Classes is bound up with his personal conduct. If he is suffering from want and misery it is because he must be vicious and sinful. Starting from this hypothesis, this school of social workers concentrates all its efforts and its resources on fostering personal virtue by adopting a programme which includes items such as temperance, gymnasium, cooperation, libraries, schools etc., which are calculated to make the individual a better and virtuous individual. In my opinion there is also another method of approach to this problem. It starts with the hypothesis that the fate of the individual is governed by his environment and the circumstances he is obliged to live under and if an individual is suffering from want and misery it is because his environment is not propitious.

Like the Negro in America he is the last to be employed in days of prosperity and the first to be fired in days of adversity. And even when he gets a foothold what are his prospects? In the Cotton Mills in Bombay and Ahmedabad he is confined to the lowest paid department where he can earn only Rs. 25 per month. More paying departments like the weaving department are permanently closed to them. The place of the boss is reserved for the caste Hindu while the Depressed Class worker must slave as his underdog no matter how senior or how efficient.

Depressed Class women working in the winding or reeling departments have come to me in hundreds complaining that the Naikins, instead of distributing the raw material to all women employees equally or in fair proportion, give all of it to the caste Hindu women and leave them in the cold.

I think it would be fit and proper if the Anti-Untouchability League were to take up this question by creating public opinion in condemnation of it and establishing Bureaus to deal with urgent cases of inequality.

Lastly I think the League should attempt to dissolve that nausea which the touchables feel towards the untouchables and which is the reason why the two sections have remained so much apart as to constitute separate and distinct entities. In my opinion the best way of achieving it is to establish closer contact between the two. Only a common cycle of participation can help people to overcome that strangeness of feeling which one has when brought into contact with the other. Nothing can do this more effectively in my opinion than the admission of the Depressed Classes to the houses of the caste Hindus as guests or servants.

The live contact thus established will familiarize both to a common and associated life and will pave the way for that unity which we are all striving after. I am sorry that many caste Hindus who have shown themselves responsive are not prepared for this.

During those ten days of the Mahatma's fast that shook the Indian world there were cases in Ville Parle and in Mahad where the caste Hindu servants had struck work because their masters had abrogated the rules of untouchability by fraternising with the Untouchables. I expected that they would end the strike and teach a lesson to the erring masses by filling the vacancies by employing Depressed Classes in their places. Instead of doing that they capitulated with the forces of orthodoxy and strengthened them. I do not know how far such fair-weather friends of the Depressed Classes would be of help to them.

People in distress can have very little consolation from the fact that they have sympathisers if those sympathisers will do nothing more than sympathise and I may as well tell the League that the Depressed Classes will never be satisfied of the bona fides of these caste Hindu sympathisers until it is proved that they are prepared to go to the same length of fighting against their own kith and kin in actual warfare if it came to that for the sake of the Depressed Classes as the Whites of the North did against their own kith and kin namely the Whites of the South for the sake of the emancipation of the Negro.

Secondly there are agencies which are already engaged in some sort of social service without any confines as to class or purpose and may be prepared to supplement their activity by taking up the work of the Anti-Untouchability League in consideration of a grant-in-aid. I am sure this hire-purchase system of work, if I may use that expression, can produce no lasting good. What is wanted in an agency is a single-minded devotion to one task only. We want bodies and organizations which have deliberately chosen to be narrow-minded in order to be enthusiastic about their cause. The work if it is to be assigned must be assigned to those who would undertake to devote themselves exclusively to the work of the Depressed Classes.

Before closing this I wish to say just this. It was Balfour, I think, who said that what could hold the British Empire together was love and not law. I think that observation applies equally well to the Hindu society. The touchables and the untouchables cannot be held together by law, certainly not by any electoral law substituting joint electorates for separate electorates. The only thing that can hold them together is love.

Outside the family, justice alone, in my opinion can open the possibility of love, and it should be the duty of the Anti-Untouchability League to see that the touchable does, or failing that, is made to do justice to the untouchable. Nothing else in my opinion can justify the project or the existence of the League." This letter was not even acknowledged by the Secretary. That not a single suggestion of mine was accepted goes without saying. Even my suggestion that the workers of the Sangh should be drawn largely from the Untouchables themselves was not accepted. Indeed when the attention of Mr. Gandhi was drawn to the fact that the Harijan Sevak Sangh had become the hive of mercenary Hindus, he defended it on the ground which are clever if not honest. He said to the deputation of the Untouchables.

The welfare work of the Untouchables is a penance which the Hindus have to do for the sin of Untouchability. The money that has been collected has been contributed by the Hindus. From both points of view the Hindus alone must run the Sangh. Neither ethics nor right would justify Untouchables in claiming a seat on the Board of the Sangh.

Not only were all my proposals rejected by Mr. Gandhi and his advisers but in framing the constitution of the Sangh, aims and objects were adopted which are quite opposed to those which I had suggested. At the meeting held in Cowasjee Jehangir Hall in Bombay on the 30th September 1932 the aims of the organization were stated to be: "Carrying propaganda against Untouchability and taking immediate steps 'to secure as early as practicable that all public wells, dharmashalas, roads, schools, crematoriums, burning ghats and all public temples be declared open to the Depressed Classes, provided that no compulsion or force shall be used and that only peaceful persuasion shall be adopted towards this end'."

But in the statement issued by Mr. G. D. Birla and Mr. A. V. Thakkar on the 3rd November, two months after its inauguration it was stated: "The League believes that reasonable persons among the Sanatanists are not much against the removal of Untouchability as such, as they are against inter-caste dinners and marriages. Since it is not the ambition of the League to undertake reforms beyond its own scope, it is desirable to make it clear that while the League will work by persuasion among the caste Hindus to remove every vestige of Untouchability, the main line of work will be constructive, such as the uplift of Depressed Classes educationally, economically and socially, which itself will go a great way to remove untouchability. With such a work even a staunch Sanatanist can have nothing but sympathy. And it is for such work mainly that the League has been established. Social reforms like the abolition of the caste system and inter-dinning are kept outside the scope of the League."

These aims and objects are described in one of the Annual Reports of the Sangh. It says: "According to its constitution the aim and object of the Society is the abolition of untouchability by reason of birth and the acquisition of equal rights of access of public temples, wells, schools and other public institutions for Harijans as enjoyed by other Hindus. The achievement of this object has led the Society to undertake work of a two-fold kind. First, the Society has to bring about such a radical change in the sentiments and opinions of Caste Hindus that they may willingly, as a matter of course, allow the enjoyment of all civic rights to Harijans. Secondly, the society has to put forth its efforts and devote its funds for the educational, economic and social uplift of Harijans."

The work done and the aims formulated when put side by side raise two questions. Firstly is this record something of which the Sangh can be proud of? Secondly is its work consistent with the aims of the Sangh? The record is very poor. It is much cry and little wool. Certainly as compared with the record of work done by the Christian Missions with which the Sangh competes, it is not a record of which the Sangh can be proud of. But this is a mere matter for sorrow and nothing more. The second question is fundamental and therefore one for anxious consideration. It is well that the Sangh undertakes to labour in the interests of the Untouchables. But its labours must be so planned that out of it will come the destruction of untouchability.

Examined in the light of this consideration what is one to say of the work that is being done by the Sangh? The Sangh is openly and without abashment supporting *separate* schools, *separate* hostels, *separate* dispensaries, and *separate* wells for the Untouchables. I should have thought that that was the surest way of perpetuating untouchability. It is surprising that Mr. Gandhi who threatened to fast unto death against separate electorates on the ground that it involved segregation of the untouchables should himself sanction a line of activity which perpetuates this segregation. In undertaking to render this social service to the Untouchables, Mr. Gandhi and his Sangh should have forgotten what the Untouchables want. What the Untouchables want is not education, but the right to be admitted to common schools.

The Untouchables do not want medical aid; what they want is the right to be admitted to the general dispensary on equal terms. The Untouchable does not want water. What he wants is the right to draw water from a common well. The Untouchables do not want their suffering to be relieved. They want their personality to be respected and their stigma removed. Once their stigma is removed their sufferings will go. This the Harijan Sevak Sangh does not seem to have realized.

The Sangh is said to be the friend of the Untouchables and the orthodox Hindu the enemy of the Untouchables. One fails to understand what the friend has done which the enemy would not do. The orthodox Hindu insists that the Untouchables shall have *separate* schools, *separate* dispensaries and *separate* wells, the Sangh says—*Thy will shall be done.* Except the fact the orthodox Hindu believes in untouchability and Harijan Sevak Sangh does not, what is the difference in practice between the friend and the foe? Under both, the untouchable is condemned to *separate* schools, *separate* hostels and *separate* wells. If this is so, it is difficult to understand why.

Mr. Gandhi and the Harijan Sevak Sangh should pick up a quarrel with the orthodox Hindu if he and his Sangh are not prepared to force the issue. Whether the Hindu Shastras recognize untouchability or not is only an academic quarrel between Mr. Gandhi and the orthodox Hindu. It can do no practical good. On the contrary I am prepared to say that it had done positive harm to the Untouchables. In the first place it has created enmity between the Untouchables and the orthodox Hindus. Before Gandhi picked up this needless quarrel the relations between the Untouchables and the Hindus were non-social.

The quarrel has made them anti-social. Secondly if there was no such quarrel, if instead of untouchability being made the issue—which Mr. Gandhi does not intend to fight it out—an appeal was made to the orthodox Hindu to remove the suffering of the Untouchables, many an orthodox Hindu I know would have come forward to help to remove the suffering. Mr. Gandhi has reaped the glory for having established the Sangh. But the Sangh has neither sought to remove untouchability nor has it helped to alleviate the sufferings of the Untouchables.

Why has the Sangh failed? My answer is quite definite. I say the Sangh has failed because of its wrong politics. It has often been said that the Harijan Sevak Sangh is a political organization. Mr. Gandhi has always resented such an allegation and repudiated it as being false. The General Secretary of the Sangh has also protested against it. To use his own words "the Sangh, though a sequel of a Political Pact, has no politics"

I do not see any reason for the resentment of Mr. Gandhi nor for the protests of his Secretary. I wish very much that the Sangh was a political organization. The untouchables have obtained a share of political power. But power which is not conscious of itself is no power. Again power which is not organized is no power. The Harijan Sevak Sangh would have been of great use if it had helped the Untouchables to organize independent political parties to fight the elections and make their political power effective. Nor can I accept the statement of Mr. Gandhi and his Secretary that the Sangh has no politics. On the contrary I insist that not only the Sangh has a definite line of politics and that line of politics is wrong because it is prejudicial to the cause of the Untouchables.

Since Mr. Gandhi does not admit that the Sangh has politics, one must go to circumstances for proof. Circumstantial proof is always better than oral testimony because as is well said man may lie but circumstances cannot. In this connection I want to rely upon a clause in the constitution of the Sangh as a piece of evidence in support of my contention. The clause relates to the means to be adopted by the Sangh for removing untouchability and for securing equal rights to the Untouchables along with the Caste Hindus. The clause reads as follows: "That no compulsion is to be used for securing rights, but that peaceful persuation is to be adopted as the *only* means." This is a basic principle of the Sangh. It has struck me as strange and I am sure it will strike all others as strange. I want to ask the question—Why has the Sangh limited itself to peaceful persuation of the caste Hindus as the one and the only means of removing Untouchability?

Most social reformers, whether religious or rational, seem to imagine that men of power will immediately check their pretentions and their exactions as soon as they have been told that their actions and attitudes are anti-social. But as Prof. Neibhur points out what these reformers overlook is an understanding of the brutal character of the behaviour of all human groups and the power of self interest and collective egoism which dominate all group relations. They also forget the fact that races, nations and classes are less moral than individuals which compose them and that justice between groups can therefore not be achieved purely by educational means. If conscience and reason can be insinuated into the resulting struggle they can only qualify, never abolish, the injustice If injustice is to be abolished it must be resisted and when injustice proceeds from collective power, whether in the form of imperialism or class domination, it must be challenged by power. A class entrenched behind its established power can never be dislodged unless power is raised against it. That is the only way of stopping the exploitation of the weak by strong.

Why has Mr. Gandhi and the Harijan Sevak Sangh limited their means of resistance to the Caste Hindu domination to peaceful persuation? Why do they not resist the injustice of the Caste Hindus by direct action? I can understand that in organising resistance to injustice, the problem is to find forms of resistance which will not destroy the meagre resources for rational and moral action which groups do possess. But there can be no difficulty on that account. Satyagraha or passive resistance has been found by Mr. Gandhi as a form of resistance which is morally beyond cavil. Why does not Mr. Gandhi ask the Sangh to launch Satyagraha by the Untouchables against the Caste Hindus for the abolition of the injustice against the Untouchables. He asked the people of India to offer Satyagraha against British Imperialism. Why does he not want to use the same means against the caste Hindus in the interests of the Untouchables?

What is the answer of Mr. Gandhi to this question? The only answer I can see is that it comes in the way of *his* politics. Mr. Gandhi must remain at the head of the nation. I wonder if life would be worthwhile to him if for some reason he ceased to be at the head of the nation. He is, I think, the most ambitious politician. I know, he regards as his rivials those whom he calls as friends. To be at the head of the nation means that he must preserve the integrity of the Congress. The Congress is ninety nine per cent composed of Hindus. How can Gandhi succeed in maintaining the integrity of the Congress if he were to direct the Sangh to carry on Satyagraha against the Hindus for the sake of the Untouchables. The Hindus would leave the Congress and the Congress would disrupt. This is detrimental to the interests of Mr. Gandhi. This is the explanation why Mr. Gandhi and the Sangh have adopted peaceful persuation as the only means of removing untouchability. It is a means which is least likely to hurt the Hindus and the Congress. Not only in big matters but even in small matters the Sangh is careful to see that the Hindus are not hurt or annoyed. I am told that in distributing scholarships for instance the Sangh makes inquiries into the political affiliations of the applicant and if it is found that the applicant belongs to a community which is against the Congress or the Hindus, he gets no aid from the funds of the Sangh.

I wonder if any one will have any doubts left that Mr. Gandhi and the Sangh in limiting themselves to peaceful persuations were controlled by political considerations of not annoying the Hindus and disrupting the Congress. This is what I meant when I said that the Sangha's politics and that its failure is due to wrong politics. I am sure I am using mild language when I describe it as wrong politics. It is treachery if the surrender of the interests of his ward by his guardian can be described as treachery.

Mr. Gandhi is often compared with Jesus Christ both by his Indian and European friends. What may be the justification for so strange a comparison? In one thing I see a complete contrast between the two. Both Jesus and Gandhi claimed to serve the lowly. This befriending attitude of both was resented by the upper classes. How did the two react? When Jesus was taunted by the Pharasees he retorted by saying—" They that are whole have no need of the physician, but they that are sick". How sharp is the contrast between this attitude of Jesus and that of Mr. Gandhi. Jesus did not worry about those who were 'whole'. Gandhi is devoted to those who are 'whole' and who are sinning at the cost of those who are sick and who are sinned against. Gandhi is no physician to the untouchables. At best a sympathiser and nothing more.

Even as a sympathizer of the Untouchables his sympathy for them is limited by two considerations. It is limited by his social aims. Secondly it is also limited by his politics. Lest this statement should be doubted, I wish to give two instances, one of each which have occurred recently. They have occurred not far from Shegaon in Central provinces where Mr. Gandhi resides.

As an instance of the first I refer to what is known in India as the Khare episode. In 1938 last there was a ministerial crisis in the Central Provinces where the ministry was a Congress Ministry. The Prime Minister Dr. Khare fell out with his colleagues. As Prime Minister he demanded the resignation of other ministers which they were not prepared to give without the consent of the Congress Parliamentary Board. But Dr. Khare forced them to give their resignations which they ultimately did. Thereupon Dr. Khare formed another Congress cabinet and filled the ministerial offices with men of his choice. In his new cabinet Dr. Khare included an untouchable as a minister. Dr. Khare's conduct in dissolving the old cabinet and forming a new one without consulting the Congress Parliamentary Board came up for investigation before the Congress Working Committee. Dr. Khare was found guilty of breach of party discipline and was deposed from his primiership. One of the accusations levelled against Dr. Khare by Mr. Gandhi was that he included an untouchable in his new Ministry.

The following is the full text of what Mr. Gandhi said to Dr. Khare on this point and reduced to writing by Dr. Khare for my sake: "Mahatmaji took me to task for including a Harijan in my second cabinet. I retorted by saying that it was a Congress programme of uplift of Harijans for which Mahatmaji fasted unto death and that I did what I could in furtherance of that programme when opportunity offered itself and I think I have done nothing wrong in doing so. Thereupon Mahatmaji charged me of doing this for my selfish ambition. I repudiated this charge saying that any selfish motive is disproved by my resignation. Then Mahatmaji said that by my action I have thrown an apple of discord among the members of that simple community and have rendered disservice to the Congress cause by throwing this temptation in their way."

That this is true and that Gandhi objected to the untouchable being included in the Cabinet is evidenced by the fact that when a new Congress Ministry was formed in Central Provinces this untouchable who functioned as a minister for a day was excluded. He should have been included as a matter of form, at least to keep up appearances. That he was not, shows that Gandhi was opposed to his inclusion on principle.

This is baffling because the Untouchable member of the Central Provinces Assembly who was chosen by Dr. Khare for ministership is a graduate, is a Congressman and is a strong party-man. Why should Mr. Gandhi have any objection to the inclusion of such a person in the Congress Cabinet. As a matter of fact if Mr. Gandhi was genuine in his professions regarding the untouchables he should have instructed all Congress Prime Ministers to include at least one untouchable in their Cabinets, if for nothing else, at least for its psychological effect upon the Untouchables and upon the Hindus. He should have done this irrespective of the party affiliations. Gandhi is not averse to congress making coalitions with other non-congress parties in provinces where it is not in a majority in order to secure offices. In such coalitions he has allowed Congress to include non-congressmen as ministers in their cabinets. If Gandhi can allow the congress to do this without losing its caste and its colour why Gandhi did not instruct the Congress Prime Ministers to include untouchables in the Congress Ministries if he wanted that when he talks about his love for the Untouchables he should be believed. But the case in the Central Provinces stand on a different footing. Here the Untouchable who was taken as a minister was a Congressman and a graduate. There could be no objection on the ground of his want of qualification or want of political faith. Why did Mr. Gandhi object to his inclusion? A crowd of Untouchables went to Shegaon to Mr. Gandhi for an explanation. Anticipating this Mr. Gandhi had started observing silence, so that no explanation could be had. Then the untouchables started Satyagraha against Mr. Gandhi for not including an Untouchable in the Central Provinces Congress Cabinet.

To escape that embarassment Mr. Gandhi left Shegaon and went on a tour to the North Western Frontier Provinces for teaching non-violence to the Pathans. I am sure Gandhi's silence on this occasion was not to commune with God. It was taken on as a convenient excuse for not being driven under the fire of cross examination to disclose his inner-most thoughts about the Untouchables. In any case we have no answer to this question from Mr. Gandhi. To my mind there can be only one answer and that even if Gandhi had opened his mouth he could give no better. That answer is that Mr. Gandhi's ideal for the untouchable is a very low ideal and that all that he cares for is that the untouchable should be touched and that if he is touched without anybody taking a bath nothing further need be done about them. If Gandhi had tried and failed he would have been excused. But how can he be excused for entertaining so low an aim? Not failure but low aim is a crime.

As an instance of the second I refer to what is known as the Bachuma incident. If the first instance is baffling this second is gruesome. How gruesome it is can be seen from the brief summary of fact which I am giving. Bachuma, a small girl, 12 years old and belonged to an untouchable family which was living in *Wardha*. One evening she was decoyed into the house of a Mohammedan who was the Sub-Inspector of Police. She was kept in his house and during the night this small child was raped by three Mohammedans, one was the Superintendent of Police himself, second a Sub-Inspector of Education and third a Lawyer. The three Mohammedans were tried in a Court of Law and two of them were sentenced to two years rigorous imprisonment and the lawyer was acquitted as the girl was not able to identify him.

The two who were convicted appealed to the High Court but the High Court rejected their appeals and confirmed their convictions and sentences. From gaol they sent to the Governor-in-Council petitions for mercy. But they were also rejected. This happened before the Congress came into office. After the Congress came into office they submitted fresh applications for mercy to the Minister-in-charge. The Minister-in-charge, who was also a Mohammedan thought that there was nothing wrong in a Mohammedan committing rape on an untouchable girl and decided to set the culprits free. He granted the application of one—that of the Inspector of Education who is now a free man and is employed on a big job in the Education Department of a Mohammedan State. He was to release the other culprit also but in the meantime the agitation against him was so great that he had to resign his office.

Every body expressed his resentment against the shameless act of the Minister but Mr. Gandhi has kept mum. So far he has not uttered a word of condemnation against this Minister. On the contrary he is even now engaged in the confabulations that are going on over the question of the reinstatement of this dismissed minister in his office which is still kept vacant. One likes to ask if Gandhi would have remained so silent and so unmoved if the little girl Bachuma who was raped by the three Mohammedans instead of being the daughter of an Untouchable had been Mr. Gandhi's own daughter. Why is Gandhi not able to make Bachuma's case his daughter's case? There are two answers. One is that Mr. Gandhi is not an untouchable. One must be born to it. Secondly Mr. Gandhi feels that by condemning the Muslim Minister for the sake of Bachuma he might destroy Hindu-Moslem Unity the maintenance of which is a fundamental creed of Congress politics. Does this now show that Mr. Gandhi's sympathies for the Untouchables are limited by his politics?

What good is a man who is not even free to sympathize according to his conscience. And now the temple entry. This is advertized as another of Mr. Gandhi's boons to the Untouchables. This question of temple entry is the outcome of the resolution passed at the public meeting of the Hindus held on 25th September 1932 which also was the originating cause which gave birth to the Harijan Sevak Sangh. This resolution mentioned some of the liabilities of the Hindus towards the Untouchables. This list included the removal of the bar against the Untouchables in respect of admissions to Hindu temples.

Although the promise of temple entry was there in the Resolution the Untouchables did not insist upon its being fulfilled forthwith. The untouchables, at any rate a vast majority of them have not been keen for temple entry. When asked by Mr. Gandhi what I thought about temple entry I gave my opinion on it in the form of a statement which was issued to the Press on February 10, 1933, and which is reproduced below:

Statement on Temple Entry

Although the controversy regarding the question of temple entry is confined to the Sanatanists and Mahatma Gandhi, the Depressed Classes have undoubtedly a very important part to play in it, in so far as their position is bound to weigh the scales one way or the other, when the issue comes for final settlement. It is, therefore necessary, that their view point should be defined and stated so as to leave no ambiguity about it.

To the Temple Entry Bill of Mr. Ranga lyer as now drafted, the Depressed Classes cannot possibly give their support. The principle of the Bill is that if a majority of Municipal and Local Board voters in the vicinity of any particular temple on a Referendum decide by a majority that the Depressed Classes shall be allowed to enter the temple, the Trustees or the Manager of that temple shall give effect to that decision. The principle is an ordinary principle of majority rule and there is nothing radical or revolutionary about the Bill and if the Sanatanists were a wise lot, they would accept it without demur.

The reasons why the Depressed Classes cannot support a Bill based upon this principle are two. One reason is that the Bill cannot hasten the day of Temple Entry for the Depressed Classes any nearer than would otherwise be the case. It is true that under the Bill the minority will not have the right to obtain an injunction against the Trustee or the Manager who throws open the temple to the Depressed Classes in accordance with the decision of the majority. But before one can draw any satisfaction from this clause and congratulate the author of the Bill, one must first of all feel assured that when the question is put to the vote there will be a majority in favour of the Temple Entry. If one is not suffering from illusions of any kind, one must accept that the hope of a majority voting in favour of Temple Entry will be rarely realised, if at all. Without doubt the majority is definitely opposed to day—a fact which is conceded by the Author of the Bill himself in his correspondence with the Shankaracharya.

What is there in the situation as created after the passing of the Bill, which can lead one to hope that the majority will act differently? I find nothing. I shall, no doubt, be reminded of the results of the Referendum with regard to the Guruvayur Temple. But I refuse to accept a referendum so overweighed as it was by the Life of Mahatma Gandhi as the normal result. In any such calculations, the life of the Mahatma must necessarily be deducted. Secondly, the Bill does not regard untouchability in temples as a sinful custom. It regards untouchability merely as a social evil not necessarily worse than social evils of other sorts.

For, it does not declare untouchability as such to be illegal. Its binding force is taken away only if a majority decides to do so. Sin and immorality cannot become tolerable because a majority is addicted to them or because the majority chooses to practice them. If untouchability is a sinful and an immoral custom, then in the view of the Depressed Classes it must be destroyed without any hesitation, even if it was acceptable to the majority. This is the way in which all customs are dealt with by Courts of Law, if they find them to be immoral and against public policy.

This is exactly what the Bill does not do. The author of the Bill takes no more serious view of the custom of untouchability than does the temperance reformer of the habit of drinking. Indeed, so much is he impressed by the assumed similarity between the two that the method he has adopted is a method which is advocated by temperance reformers to eradicate the evil habit of drinking, namely by local option.

One cannot feel much grateful to a friend of the Depressed Classes who holds untouchability to be no worse than drinking. If Mr. Ranga lyer had not forgotten that only a few months ago Mahatma Gandhi had prepared himself to fast unto death if untouchability was not removed, he would have taken a more serious view of this curse and proposed a most thorough-going reform to ensure its removal lock, stock and barrel. Whatever its shortcomings may be from the standpoint of efficacy, the least that the Depressed Classes could expect is for the Bill to recognise the principle that untouchability is a sin. I really cannot understand how the Bill satisfies Mahatma Gandhi who has been insisting that untouchability is a sin.

It certainly does not satisfy the Depressed Classes. The question whether this particular Bill is good or bad, sufficient or insufficient, is a subsidiary question. The main question is; do the Depressed Classes desire Temple Entry or do they not? This main question is being viewed by the Depressed Classes by two points of view.

One is the materialistic point of view. Starting from it, the Depressed Classes think that the surest way for their elevation lies in higher education, higher employment and better ways of earning a living. Once they become well placed in the scale of social life they would become respectable and once they become respectable the religious outlook of the orthodox towards them is sure to undergo change, and even if this did not happen it can do no injury to their material interest. Proceeding on these lines the Depressed Classes say that they will not spend their resources in such an empty thing as Temple Entry.

There is also another reason why they do not care to fight for it. That argument is the argument of ' self respect '. Not very long ago there used to be boards in club doors and other social resorts maintained by Europeans in India, which said *'Dogs and Indians not allowed"*. The Temples of the Hindus carry similar boards today, the only difference is that the boards on the Hindu temples practically say *"All Hindus and all animals including dogs are admitted only Untouchables not admitted"*. The situation in both cases is on a parity. But the Hindus never begged for admission in those places from which the Europeans in their arrogance had excluded them. Why should an untouchable beg for admission in a place from which he has been excluded by the arrogance of the Hindus? This is the reasoning of the Depressed Class man who is interested in his material welfare.

He is prepared to say to the Hindus, "To open or not to open your temples is a question for you to consider, and not for me to agitate. If you think, it is bad manners not to respect the sacredness of human personality, open your temples and be a gentleman. If you rather be a Hindu than be gentleman, then shut the doors and damn yourself, for I don't care to come." "I found it necessary to put the argument in this form, because I want to disabuse the minds of men like Pandit Madan Mohan Malaviya of their belief that the Depressed Classes are looking forward expectantly for their patronage.

The second point of view is the spiritual one. As religiously minded people, do the Depressed Classes desire temple entry or do they not? That is the question. From the spiritual point of view, they are not indifferent to temple entry as they would be, if the material point of view alone were to prevail. But their final answer must depend upon the reply which Mahatma Gandhi and the Hindus give to the following question:

What is the drive behind this offer of temple entry? Is temple entry to be the final goal of the advancement in the social status of the Depressed Classes in the Hindu fold? Or is it only the first step and if it is the first step, what is the ultimate goal? Temple entry as a final goal the Depressed Classes can never support. Indeed they will not only reject it, but they would then regard themselves as rejected by Hindu Society and free to find their own destiny elsewhere. On the other hand, if it is only to be a first step in the direction, they may be inclined to support it. The position would then be analogous to what is happening in the politics of India today. All Indians have claimed Dominion Status for India.

The actual constitution will fall short of Dominion Status and many Indians will accept it. Why? The answer is that as the goal is defined, it does not matter much if it is to be reached by steps and not in one jump. But if the British had not accepted the goal of Dominion Status, no one would have accepted the partial reforms which many are now prepared to accept. In the same way if Mahatma Gandhi and the reformers were to proclaim what the goal which they set before themselves is for the advancement of the social status of the Depressed Classes in the Hindu fold, it would be easier for the Depressed Classes to define their attitude towards temple entry.

The goal of the Depressed Classes might as well be stated here for the information and consideration of all concerned. What the Depressed Classes want is a religion, which will give them equality of social status. To prevent any misunderstanding, I would like to elaborate the point by drawing a distinction between social evils which are the results of secular causes and social evils which are founded upon the doctrines of religion. Social evils can have no justification whatsoever in a civilised society. But nothing can be more odious and vile than that admitted social evils should be sought to be justified on the ground of religion.

The Depressed Classes may not be able to overthrow inequities to which they are being subjected. But they have made up their mind not to tolerate a religion that will lend its support to the continuance of these inequities. If the Hindu religion is to be their religion then it must become a religion of Social Equality. The mere amendment of Hindu religious code by the mere inclusion in it of a provision to permit temple entry for all, cannot make it a religion of Equality of social status.

All that it can do is to recognise them as nationals and not aliens, if I may use these terms which have become so familiar in politics. But that cannot mean that they would thereby reach a position where they would be free and equal without being above or below any one else, for the simple reason that the Hindu religion does not recognise the principle of equality of social status; on the other hand fosters inequality by insisting upon grading people as Brahmins, Kashatriyas, Vaishyas and Sudras which now stand towards one another in an ascending scale of hatred and descending scale of contempt. If the Hindu religion is to be a religion of social equality then an amendment of its code to provide temple entry is not enough.

What is required is to purge it of the doctrine of Chaturvarna. That is the root cause of all inequality and also the parent of the caste system and untouchability which are merely forms of inequality. Unless it is done not only will the Depressed Classes reject temple entry, they will also reject the Hindu faith. Chaturvarna and the caste system are incompitable with the self respect of the Depressed Classes. So long it continues to be the cardinal doctrine, the Depressed Classes must continue to be looked upon as low. For the Depressed Classes to say that they are Hindus is to admit their inferiority of status by their own mouth.

They can call themselves as Hindus only when the theory of Chaturvarna and caste system is abandoned and expunged from the Shastras. Do the Mahatma and the Hindu reformers accept this as their goal and will they show the courage to work for it? I shall look forward to their pronouncements on this issue as I have stated it with great concern. But whether Mahatma Gandhi and the Hindus are prepared for this or not, let it be known once for all that nothing short of this will satisfy the Depressed Classes and make them accept temple entry. To accept temple entry and be content with it, is to temporize with evil and barter away the sacredness of human personality that dwells in them.

There is however one argument which Mahatma Gandhi and the reforming Hindus may advance against the position I have taken. They may say, "acceptance by the Depressed Classes of Temple entry now, will not prevent them from agitating hereafter for the abolition of Chaturvarna and caste ". If that is their goal, I like to meet this argument right at this stage with a view to clinch the issue and clear the road for future developments. My reply is that it is true that my right to agitate for the abolition of Chaturvarna and caste system will not be lost, if I accept Temple Entry now. But the question is on what side will Mahatma Gandhi be at the time when the question is put. If he will be in the camp of my opponents I must tell him that I cannot be in his camp now. If he will be in my camp, he ought to be in it now".

My friend Dewan Bahadur R. Srinivasan expressed himself almost in the same terms on the question of temple entry. He said: "When a Depressed Classes member is permitted to enter into the Caste Hindu temples he would not be taken into any one of the four castes, but treated as man of fifth or the last or the lowest caste, a stigma worse than the one to be called an untouchable. At the same time he would be subjected to so many caste restrictions and humiliations. The Depressed Classes shun the one who enters like that and exclude him as Casteman. The crores of Depressed Classes would not submit to caste restrictions. They will be divided into sections if they do.

Temple entry cannot be forced by law. The village castemen openly or indirectly defy the law. To the village Depressed Classman it would be like a scrap of paper on which word "sugar" was written and placed in his hands for him to taste. The above facts are placed before the public in time to save confusion and disturbance in the country."

But Mr. Gandhi felt otherwise that securing temple entry to the untouchables was a liability of the Hindus which ought to be liquidated first. Accordingly immediately after the Poona Pact he started a campaign among the Hindus for opening the doors of their temples to the Untouchables.

How far has Mr. Gandhi succeeded in this matter is a question that may legitimately be asked. But it is difficult to know the truth. As a result of the fast, many temples were reported to have been thrown open by the Hindus to the Untouchables. How far this was true and how far it is a part of lying propaganda which the Congressman is so good at it is difficult to say. That many of the temples that were opened as a sequel to the fast were purified and closed to the Untouchables is beyond dispute. Again the opening of a temple may be quite a meaningless act. There are hundreds and thousands of temples in which there is no worship. They are occupied by only donkies. Instances of such temples can be seen at places of pilgrimage such as Nasik and Wai. If such a temple is declared to be open it is not only a meaningless act but it is an insult to the Untouchables. Again a temple may be opened to the Untouchable. But if it is abandoned by the Caste Hindus as a place of worship it cannot be said that it is open in the sense that they are welcomed to it by the Hindus. There is yet another possibility which must be taken into account in arriving at the truth.

A temple may be open to the untouchables in the same sense as the Ritz Hotel in London is open to all. We however know that the Ritz Hotel is not in fact open to all. It is open only to those who can afford. In the same way a temple may be open to the untouchables yet in fact it is open only to those untouchables who can afford to enter. If the cost of temple entry is assault or social boycott then the cost will be prohibitive and the temple though nominally open is really closed. Assault and social boycott are a matter of course with the Hindus and it would not be too much to assume that in some case the Hindus would resort to such means to prohibit the Untouchables who dare to enter a temple which is declared to be open to them. If the case is one like this then it is a fraud.

Which of the two classes of cases are more numerous, it is difficult to say accurately. But a guess may be made on the basis of certain facts. There are two classes of Hindus now in India—the orthodox Hindus who care more for religion than for politics and the Congress Hindus who care more for politics and less for religion. The former who have no political ends to subserve can be honest *i.e.* true to their convictions however wrong they may be. The latter who have to serve political ends cannot always take an honest view but are prone to adopt dishonest ways.

The first method of abandonment though honest brings discredit upon the Hindu community in the eyes of the world and is therefore politically unsuitable. The second method of opening the temple nominally and closing it really by Hindus is politically highly advantageous. It has the merit of a system which shows to the world that credit is opened and which clandestinely but without the world knowing prevents its being drawn upon by the person in whose favour it is declared to be opened. The Congress Hindus are more numerous than the orthodox Hindus. That being the case I should think that the second classes of cases must be more numerous than the first.

That genuine cases of opening of temples are very few and that most of the published reports of opening of temples is just false propaganda is clear from the fate of the Temple Entry Bill of Mr. Ranga lyer brought by him in the Central Legislature in 1934. Of that Bill I will speak of at a later occasion.

With this I would have closed this discussion of the subject. But Mr. Gandhi insists that a spiritual awakening has taken place among the Hindus and relies upon the Temple Entry Proclamation of Travancore. I am therefore obliged to deal with this claim. The success of temple entry cannot be determined by the number of temples opened. It can be determined only by reference to the motive with which it is done. Is the motive spiritual? That can be the only test. Now I say that temple entry is not a spiritual act. It is a political manoeuvre. Is Mr. Gandhi acting from spiritual motive? In appealing to the Hindus Mr. Gandhi said: "I have adressed this appeal to you, which proceeds out of my soul's agony. I ask you to share that agony and shame with me and cooperate with me, for I have no other end to serve than to see *Sanatana Dharma* revivified and lived in its reality in the lives of millions who at present seem to me to deny it. "This soul's agony was born after the Poona Pact. What did Mr. Gandhi think of the Problem of Temple Entry before the Poona Pact? Before the Poona Pact Mr. Gandhi was of different opinion. That opinion was expressed not very long before the Poona Pact and not long before this appeal was addressed to the Hindus from which I have....... It was expressed in Gandhi Shikshan, Vol. II, p. 132. Mr. Gandhi then held the opinion that separate temples should be built for the use of Untouchables.

Mr. Gandhi said: "How is it possible that the Antyajas (Untouchables) should have the right to enter all the existing temples? As long as the law of caste and ashram has the chief place in Hindu Religion, to say that every Hindu can enter every temple is a thing that is not possible today."

It is obvious that Temple Entry is not original with him and therefore not spiritual. The agony is caused by the grave and sudden provocation brought about by the demand of the untouchables for separate electorates. Mr. Gandhi was afraid of the principle underlying separate electorates. He felt that this principle may be extended and may ultimately lead to separation and cessation of the Untouchables from the Hindu fold. It was to counter this move that he changed his opinion and started the temple entry move. The motive of Mr. Gandhi is political and there is nothing spiritual about it.

I do not wish this conclusion to rest merely upon this evidence of change of front on the part of Mr. Gandhi. There is abundance of other evidence in support of it. I will refer first to the Guruvayur Temple Satyagraha which was started by a caste Hindu by name Kellappan to secure entry into the Guruvayur Temple for the Untouchables. A few facts regarding this episode may be interesting. The point to note is the attitude that Mr. Gandhi finally adopted in this matter when he was challenged by the leaders of the orthodox Hindus. Mr. Gandhi became ready for a compromise with the orthodox. The terms of the compromise were as follows. I give them in Mr. Gandhi's own words as reported in the papers. "During certain hours of the day the Guruvayur Temple should be thrown open to the Harijans and other Hindus, who have no objection to the presence of the Harijans and during certain other hours it should be reserved for those, who have scruples against the entry of the Harijans. There should be no difficulty whatsoever in the acceptance of this suggestion, seeing that in connection with the Krithikai Ekadashi festival in Guruvayur, the Harijans are allowed to enter side by side with the Hindus and then the temple or the idol undergoes purification."

Asked if his suggestion was that the temple might undergo purification daily after the entry of the Harijans, Mr. Gandhi replied: "Personally, I am opposed to purification at all. But if that would satisfy the conscience of the objectors I would personally in this case, raise no objection to purification. If purification has any value, then there are so many possibilities of daily defilement from a variety of causes referred to in various texts that there should be a daily purification, whether the Harijans are allowed to enter or not."

This attitude is not spiritual. It is purely commercial. This is almost admitted by Mr. Gandhi. Asked if the compromise suggested by him did not still maintain a distinction between the Untouchables and the Caste Hindus Mr. Gandhi is reported to have said: "The Harijans' attitude should be this, 'if there is a person who objects to my presence, I would like to respect his objection so long as he (the objector) does not deprive me of the right that belongs to me and so long as I am permitted to have my legitimate share of the days of offering worship side by side with those, who have no objection to my presence, I would be satisfied'." I do not know if any self respecting Untouchable would adopt this attitude of Mr. Gandhi. On these terms even dogs and cats are admitted in all temples when there are no human beings present in them. To divide the House of God in time or in space for worship for reconciling the rival claims of two opposing classes is in itself a quaint or grotesque idea. Mr. Gandhi evidently forgot that worshipping in the same temple is quite different from worshipping in common. Temple entry if it is to be spiritual must mean the latter. The former accepts that the presence of one class is repugnant to the other and proceeds to reconcile the interests of the two. The latter presupposes that there is no repugnance between the two classes and that they accept the common denominator of humanity as being present in both.

This shows that Mr. Gandhi is least motivated by spiritual considerations. He is in a hurry to bring the Untouchables within the Hindu stables so as to prevent their running away. Another piece of evidence which goes to disprove Mr. Gandhi's claim that he is acting from spiritual consideration is furnished by his conduct with reference to Mr. Ranga lyer's *'Temple Entry Bill'*. It shows that the soul's agony is only a picturesque phrase and not a fact.

Some history of this Bill is necessary to understand the tragedy which ultimately befell Mr. Ranga lyer the author of it. Since the new constitution came into operation two Acts have been passed in two Provinces by the Congress Governments. One in Bombay and another in Madras. There is no substance in the Acts. They do not declare the Temples to be open. They permit the Trustees of the Temples under their management if they desire and as there is nothing to compel the Trustees to do so the Acts are just scraps of paper and nothing more. But the Madras Act has a history which is somewhat puzzling. The Madras Prime Minister who got the Act passed is Mr. Rajagopalachariar. He occupies a very high place in the Congress, so high indeed that he is called Deputy

Mahatma. If one can solve the puzzle the solution will reveal the mind of the Author and therefore of Mr. Gandhi who is the living spirit behind all this.

Let me turn to the Travancore Temple Entry. The proclamation of 12th November 1936 issued by the Maharaja opening the Temple open to the Untouchables is a gorgeous document. It reads as follows: "Profoundly convinced of the truth and validity of our religion, believing that it is based on divine guidance and on all-comprehending toleration, knowing that in its practice it has throughout the centuries adapted itself to the need of the changing times, solicitous that none of our Hindu subjects should, by reason of birth, caste or community, be denied the consolation and solace of the Hindu faith, we have decided and hereby declare, ordain and command that, subject to such rules and conditions as may be laid down and imposed by us for preserving their proper atmosphere and maintaining their rituals and observances, there should henceforth be no restriction placed on any Hindu by birth or religion on entering or worshipping at temples controlled by us and our Government." What spirituality underlies this proclamation?

The proclamation was issued by the Maharaja of Travancore in his name. But the real active force behind the scene was the Prime Minister Sir C. P. Ramaswami Iyer. It is his motives that we must understand. In 1933 Sir C. P. R. Iyer was also the Prime Minister of Travancore. In 1933 Mr. Gandhi was fighting to get the Guruvayur Temple opened to all Untouchables. Among the many who took part in the controversy over the issue of Temple entry was Sir C. P. Ramaswami Iyer. No body seems now to remember this fact. But it is important to recall it because it helps us to understand the motives which prompted him to press the Maharaja to issue this proclamation. What attitude did Sir C. P. Ramawasmi Iyer have regarding this issue in 1933? It will be clear from the following statement which he issued to the press: "Personally I do not observe caste rules. I realise there are strong, though not very articulate, feelings in this matter in the minds of men who believe that the present system of temple worship and its details are based on divine ordinances.

The problem can be permanently solved only by a process of mutual adjustment and by the awakening of religious and social leaders of Hindu society to the realities of the present situation and to the need for preserving the solidarity of the Hindu community. "Shock tactics will not answer the purpose and direct action will be even more fatal in this sphere than in the political. I have the misfortune to differ from Mr. Gandhi when he says that the problem of temple entry can be divorced from such topics as interdining and I agree with Dr. Ambedkar that the social and economic uplift of the Depressed Classes should be our immediate and urgent programme."

This statement shows that in 1933 spiritual considerations did not move Sir C. P. Ramaswami lyer. Spiritual considerations have become operative after 1933. Is there any particular reason why these spiritual considerations should have been thought of in 1936? This question can be answered only if one bears in mind the fact that in 1936 there was held in Travancore a Conference of the Yezawa Community to consider the issue of conversion which was raised by me at Yeola in 1935. The Yezawas are an untouchable community spread over Malabar. It is an educated community and economically quite strong. It is also a vocal community and has been carrying on agitation in the state for social, religious and political rights. The Yezawas form a very large community. The cessation of so large a community would be a deathknell to the Hindus and the Conference had made the danger real as well as immediate. It was this which brought about a change in the attitude of Sir C. P. Ramaswami lyer. The spiritual considerations are just an excuse. They did not form the motives.

How far did this Proclamation change facts and how far it has remained a show? It is not possible to get real facts as they exist in Travancore. In the course of the discussion on the *Malabar Temple Entry Bill* in the Madras Legislative Assembly certain facts relating to Travancore were mentioned by Sir T. Pannirselvam, which if true would show that the whole thing is hollow. Sir T. Pannirselvam said: "One of the arguments advanced by the Premier in support of the measure was that temples in Travancore had been thrown open to the "Untouchables". A Maharaja vested with autocratic powers did so by an order. But how was it working there? From representations received, he was led to believe that after the first flush of enthusiasm, Harijans had left off going to temples, and people who used to worship previously before Harijans were allowed to enter the temples, had stopped worshipping in temples. He would ask the Government to tell them if the measure was really a success in Travancore."

On the third reading of the Bill, Sir T. Pannirselvam made a statement which must have come as a surprize to many. He said: "He wanted to know whether it was a fact that the private temples of the Senior Maharani were excluded from the proclamation. What was the reason for it? Again during the celebration of the marriage of the daughter of the Senior Maharani it was found necessary, so he was told, to perform purificatory ceremony of the temple. If such a purification of temples took place, what happened to the proclamation?"

These facts were not challenged by the Prime Minister. Evidently they cannot be challenged. If they are incontrovertible then the less said about the Malabar Temple Entry proclamation as a spiritual testament the better. It would not be proper to close this discussion without adverting to the fear which some Untouchables entertain regarding this Temple entry movement. It is just a movement of social reform or is it a strategy?

The special privileges which the Untouchables have got in the matter of politics, in the matter of education, in the matter of services are founded upon the fact that they are Untouchables. If they cease to be Untouchables their claim to these special privileges could at once be challenged. If untouchability goes then they would be just poor and backward. But as poor and backward they would not be entitled to any special privileges which they have as untouchables. What is the plan of these protagonists of Temle Entry? Is it just to open temples or aim is ultimately to take away the privileges? This fear is lurking in the minds of many a thinking Untouchables. That the fear is a real fear is clear from what is happening in Travancore itself. A correspondent of mine who represents the All Travancore Pulayar Cheramar Aykia Maha Sangham writes to me as follows in a letter, dated 24th November 1938. I give below the full text of the letter sent by him to me.

I have unaffected pleasure to draw your attention to the following facts for obtaining the valuable advice from you. Being the leader of a Harijan Community of the Travancore State, I think, it is my paramount duty to suggest you definitely all the grievances that the Harijans of this state are enduring.

1. The Temple Entry Proclamation issued by the H. H. The Maharaja is indeed a boon to Harijans; but the Harijans are enjoying all the other social disabilities except the temple entry. The proclamation is a check to the further concessions to us. The Government do not take any step for the amelioration of the Harijans.
2. Among 15 lakhs of Harijans, there are a few graduates, half a dozen undergraduates and 50 school finals and more than two hundred vernacular certificate holders. Though the Government have appointed a Public Service Commission, appointments of the Harijans are very few. All the appointments are given to Savarnas. If a Harijan is appointed it will be for one week or two weeks. According to the rules of the recruitment in Public Service the applicant is allowed to apply only after a year again, while a Savarna will be appointed for a year or more. When the list of the appointments is brought before the assembly, the number of appointments will be equal to the communal representation; but the duration of the post of all the Harijans will be equal to one Savarna. This kind of fraud is associating with the officials. Thus the public service is a common property of the Savarnas. No Harijan is benefitted by it.
3. There was a proclamation from H. H. the Maharaja, a few years ago that three acres of ground should be given to each Harijan to live in; but the Officials are Savarnas who are always unwilling to carry out the proclamation. Even though the Government is willing to grant large extent of ground for pasturing near Towns not a piece of the ground is given to the Harijans. The Harijans are still living in the compounds of the Savarnas and are undergoing manyfold difficulties.

Though large extents of ground lay in "Reserve", the applications of the Harijans for granting grounds are not at all regarded with importance or listen to. The most parts of the lands are benefited by the Savarnas.

4. The Government nominates every year of the election of members of the Assembly one member from each Harijan Community. Though they are elected to present the grievances of the Harijans before the Assembly, they are found to be the machinery of the Government *viz.*, the toys of the Savarna officers, who are benefited by them. Thus the grievances of the Harijans cannot be redressed any way.

5. All the Harijans of Travancore are labourers in the fields and compounds. They are the servants of the Savarnas who behave them as beasts—no body to look after or protection—every Harijan gets only 2 chs (one anna) as the wage in the most parts of the State. The social disabilities are the same to them even after the temple entry. The workers in the factories in various parts of the State of Travancore and the Officers of the State are all Savarnas and they are at present agitating for responsible Government. Now the Harijans are demanding jobs in Government and in factories but the agitation in Travancore is a Savarna agitation by which the Savarnas are making arrangements to get rid of Harijans in Public Service and factories. They plead for higher salaries and more privileges. They pay the least care to the Harijan labourers while the people of Travancore are maddened with the agitation of the workers in the factories. The standard of salary of Harijan worker is very low while the standard of a factory worker is thrice the former.

6. Due to starvation and proper means of livelihood the heads of the children of Harijans are heated as a result of which they are likely to fail in school. Before proclamation the duration of concession in high schools was for 6 years, now, it has reduced to three years by which a good number of students stopped their education after their failure.

7. There is a department for the Depressed Classes and the head of which is Mr. C. O. Damodaran (the protector of the Backward Communities). Though every year a big amount is granted for the expenditure, at the end of the year, 2/3 of the sum is lapsed by its sagacity. He is used to submit reports to the Government that there is no way of spending the amount. 95 per cent of the sum allotted for the Depressed Class is spent as the salary of the officials who are always Savarna and 5 per cent is benefited. Now the Government is going to make some colonies in three parts of Travancore. The officers are Savarnas. This scheme is, in my opinion, not a success for the Government do not pay greater to it. I regret that Travancore Government spends one anna for the Harijan cause, while Cochin State spends a rupee for the same.

The majority of the subjects of Travancore are now agitating strongly for Responsible Government under an organization "The State Congress". The leaders of this popular organization belong to the four major communities of the State namely, the Nair, Mohammedan, Christian and Ezhava community.

The President of the State Congress Mr. Thanu Pillai issued a statement in which he stressed that special concessions would be given to the Depressed Class. All the leaders of the Depressed Class have been waiting for a time to see the attitude of the State Congress. Now we come to understand that there is no reality in the promises of these leaders.

Now I am sure that the leaders have neglected the cause of the Depressed Class. The State Congress was started on the principles of nationalism and now it has become an institution of communalism.

Communal spirit is now working among the leaders. In every public speech, statement or article, the leaders mention only these four major communities, while they have no thought on us. I fear, if this is the case of the leaders of the political agitation of Travancore, the situation of the Depressed Class will be more deplorable when the Responsible Government is achieved, for the entire possession of the Government will be then within the clutches of the above mentioned communities and the Depressed Classes' rights and privileges will be devoured by the former.

In the meetings of the working committee of the State Congress 2/3 of the time had devoted in discussion concerning the strike of the Alleppey Coir Factories; but nothing was mentioned in the meeting about the Harijan workers who are undergoing manyfold difficulties. The workers in Factories are Savarnas and the agitation for obtaining Responsible Government is a kind of anti Harijan movement. The motive of every leader of the State Congress is to improve the situation (circumstance) of the Savarna. The leaders of the major communities have some mercinary attitude who are going to sacrifice the Depressed Class for their progress.

These are the conditions of the Depressed Class of the State. What are the ways by which we have to establish our rights in the State? I humbly request you to be good enough to render me your advice at this occasion. I am awaiting for the reply. Excuse me for the trouble.

If the plan of Temple Entry is ultimately to deprive the Untouchables of their statutory rights then the movement is not only not spiritual but it is positively mischievous and it would be the duty of all honest people to warn the Untouchables, *"Beware of Gandhi".*

A Warning to the Untouchables

Revolt and rebellion against the Established order is a natural part of the history of the poor in all countries of the world. A student of their history cannot but be struck by the thought entertained by them, of the way victory would come. In the theological age, the poor lived by the hope that spiritual forces would ultimately make the meek inherit the earth. In the secular age, otherwise called modern times, the poor live by the hope that the forces of historical materialism will automatically rob the strong of their strength and make the weak take their place.

In the light of this psychology, when one begins to think of the Untouchables in their role of rebels against the Hindu

Social Order one feels like congratulating them on their realization that neither spiritual forces nor historical forces are going to bring the millennium. They know full well that if the Hindu Social Order is to fall to the ground, it can happen only under two conditions. Firstly, the social order must be subjected to constant fire. Secondly, they can't subject it to constant fire unless they are independent of the Hindus in thought and in action. That is why the Untouchables are insistent upon separate electorates and separate settlements.

The Hindus on the other hand are telling the Untouchables to depend upon the Hindus for their emancipation. The Untouchables are told that the general spread of education will make the Hindus act in a rational manner. The Untouchables are told that the constant preaching of reformers against Untouchability is bound to bring about a moral transformation of the Hindus and the quickening of his conscience. The Untouchables should therefore rely on the good will and sense of duty of the Hindus. No Untouchable believes in this facile proposition. If there are any who do, they are hypocrites who are prepared to agree to whatever the Hindus have to say in order that by their grace they may be put in places reserved for the Untouchables. They are a predatory band of Untouchables who are out to feather their nests by any means open to them. The Untouchables are not deceived by such false propaganda and false hopes. It is therefore unnecessary to comment on it. At the same time, the propaganda is so alluring that it may mislead the unwary Untouchables into being ensnared by it. A warning to the Untouchables is therefore necessary.

Two agencies are generally relied upon by the social idealists for producing social justice. One is reason, the other is religion.The rationalist who uphold the mission of reason believe that injustice could be eliminated by the increasing power of intelligence. In the mediaeval age social injustice and superstition were intimately related to each other. It was natural for the rationalists to believe that the elimination of superstition must result in the abolition of injustice. This belief was encouraged by the results. Today it has become the creed of the educationists, philosophers, psychologists and social scientists who believe that universal education and the development of printing and press would result in an ideal society, in which every individual would be so enlightened that there would be no place for social injustice.

History, whether Indian or European, gives no unqualified support to this dogma. In Europe, the old traditions and superstitions which seemed to the eighteenth century to be the very root of injustice, have been eliminated. Yet social injustice has been rampant and has been growing ever and anon. In India itself, the whole Brahmin community is educated, man, woman and child. How many Brahmins are free from their belief in untouchability? How many have come forward to undertake a crusade against untouchability? How many are prepared to stand by the side of the Untouchables in their fight against injustice? In short, how many are prepared to make the cause of the Untouchables their own cause? The number will be appallingly small.

Why does reason fail to bring about social justice? The answer is that reason works so long as it does not come into conflict with one's vested interest. Where it comes into conflict with vested interests, it fails. Many Hindus have a vested interest in untouchability. That, vested interest may take the shape of feeling of social superiority or it may take the shape of economic exploitation such as forced labour or cheap labour, the fact remains that Hindus have a vested interest in untouchability. It is only natural that vested interest should not yield to the dictates of reason. The Untouchables should therefore know that there are limits to what reason can do.

The religious moralists who believe in the efficacy of religion urge that the moral insight which religion plants in man whereby it makes him conscious of the sinfulness of his preoccupation with self and thereby of the duty to do justice to his fellows. Nobody can deny that this is the function of religion and to some extent religion may succeed in this mission. But here again there are limits to what religion can do. Religion can help to produce justice within a community. Religion cannot produce justice between communities. At any rate, religion has failed to produce justice between Negroes and Whites, in the United States. It has failed to produce justice between Germans and

French and between them and the other nations. The call of nation and the call of community has proved more powerful than the call of religion for justice.

The Untouchables should bear in mind two things. Firstly, that it is futile to expect the Hindu religion to perform the mission of bringing about social justice. Such a task may be performed by Islam, Christianity, or Buddhism. The Hindu religion is itself the embodiment of inequity and injustice to the Untouchables. For it, to preach the gospel of justice is to go against its own being. To hope for this is to hope for a miracle. Secondly, assuming that this was a task which Hinduism was fitted to perform, it would be impossible for it to perform. The social barrier between them and the Hindus is much greater than the barrier between the Hindus and their men. Religion, however efficacious it may be within a community or a nation, is quite powerless to break these barriers and (make) them one whole.

Apart from these agencies of reason and religion the Untouchables are asked to trust the enlightened self-interests of the Hindu privileged classes and the fraternity of the Hindu proletarian. As to the privileged classes it be wrong to depend upon for anything more than their agreeing to be benevolent despots. They have their own class interests and they cannot be expected to sacrifice them for general interests or universal values. On the other hand, their constant endeavour is to identify their class interests with general interests and to assume that their privileges are the just payments with which society rewards specially useful and meritorious functions. They are a poor company to the Untouchables as the Untouchables have found in their conflict with the Hindus.

For Untouchables to expect to gain help from the Hindu proletariat is also a vain hope. The appeal of the Indian Communists to the Untouchables for solidarity with the Hindu proletariat is no doubt based on the assumption that the proletarian does not desire advantages for himself which he is not willing to share with others. Is this true? Even in Europe the proletarian are not a uniform class. It is marked by class composition, the higher and the lower. This is reflected in their attitudes towards social change, the higher are reformist and the lower are revolutionary. The assumption therefore is riot true. So far as India is concerned it is positively false. There is very little for a common front. Socially, there is bound to be antagonism between them. Economically, there cannot be much room for alliance.

What must the Untouchables strive for? Two things they must strive for is education and spread of knowledge. The power of the privileged classes rests upon lies which are sedulously propagated among the masses. No resistance to power is possible while the sanctioning lies, which justify that power are accepted as valid. While the lie which is the first and the chief line of defence remains unbroken there can be no revolt. Before any injustice, any abuse or oppression can be resisted, the lie upon which it is founded must be unmasked, must be clearly recognized for what it is. This can happen only with education.

The second thing they must strive for is power. It must not be forgotten that there is a real conflict of interests between the Hindus and the Untouchables and that while reason may mitigate the conflict it can never obviate the necessity of such a conflict. What makes one interest dominant over another is power. That being so, power is needed to destroy power. There may be the problem of how to make the use of power ethical, but there can be no question that without power on one side it is not possible to destroy power on the other side. Power is either economic or political. Military power is no power today. Because it is not free power. The economic power of the working class is the power inherent in the strike. The Untouchables as a part of the working class can have no other economic power. As it is, this power is not adequate for the defence of the interests of the working class. It is maimed by legislation and made subject to injunctions, arbitrations, martial law and use of troops. Much more inadequate is the Untouchables' power to strike.

The Untouchable is therefore under an absolute necessity of acquiring political power as much as possible. Having regard to his increasingly inadequate power in social and economic terms the Untouchable can never acquire too much political power. Whatever degree of political power he acquires, it will always be too little having regard to the vast amount of social, economic and political power of the Hindus.

The Untouchable must remember that his political power, no matter how large, will be of no use if he depends for representation in the Legislature on Hindus whose political life is rested in economic and social interests which are directly opposed to those of the Untouchables.

Who were the Sudras?

In the present stage of the literature on the subject, a book on the Sudras cannot be regarded as a superfluity. Nor can it be said to deal with a trivial problem. The general proposition that the social organization of the Indo-Aryans was based on the theory of *Chaturvarnya* and that *Chaturvarnya* means division of society into four classes—*Brahmins* (priests), *Kshatriyas (soldiers), Vaishyas* (traders) and *Sudras* (menials) does not convey any idea of the real nature of the problem of the Sudras nor of its magnitude.

Chaturvarnya would have been a very innocent principle if it meant no more than mere division of society into four classes. Unfortunately, more than this is involved in the theory of *Chaturvarnya.* Besides dividing society into four orders, the theory goes further and makes the principle of graded inequality, the basis for determining the terms of associated life as between the four *Varnas.* Again, the system of graded inequality is not merely notional.

It is legal and penal. Under the system of *Chaturvarnya,* the *Sudra* is not only placed at the bottom of the gradation but he is subjected to inunumerable ignominies and disabilities so as to prevent him from rising above the condition fixed for him by law. Indeed until the fifth *Varna* of the Untouchables came into being, the *Sudras* were in the eyes of the Hindus the lowest of the low. This shows the nature of what might be called the problem of the *Sudras.* If people have no idea of the magnitude of the problem it is because they have not cared to know what the population of the *Sudras* is. Unfortunately, the census does not show their population separately. But there is no doubt that excluding the Untouchables the *Sudras* form about 75 to 80 per cent of the population of Hindus. A treatise which deals with so vast a population cannot be considered to be dealing with a trivial problem.

The book deals with the *Sudras* in the Indo-Aryan Society. There is a view that an inquiry into these questions is of no present-day moment. It is said by no less a person than Mr. Sherring in his *Hindu Tribes and Castes* that: "Whether the Sudras were Aryans, or aboriginal inhabitants of India, or tribes produced by the union of the one with the other, is of little practical moment. They were at an early period placed in a class by themselves, and received the fourth or last degree of rank, yet at a considerable distance from the three superior castes.

Even though it be admitted that at the outset they were not Aryans, still, from their extensive intermarriages with the three Aryan Castes, they have become so far Aryanized that, in some instances as already shown, they have gained more than they have lost, and certain tribes now designated as Sudras are in reality more Brahmins and Kshatriyas than anything else. In short, they have become as much absorbed in other races the cletic tribes of England have become absorbed in the Anglo-Saxon race; and their own separate individuality, if they ever had any, has completely vanished."

This view is based on two errors. Firstly, the present-day Sudras are a collection of castes drawn from heterogeneous stocks and are racially different from the original Sudras of the Indo-Aryan society. Secondly, in the case of Sudras the centre of interest is not the Sudras as a people but the legal system of pains and penalties to which they are subjected. The system of pains and penalties was no doubt originally devised by the Brahmins to deal with the Sudras of the Indo-Aryan society, who have ceased to exist as a distinct, separate, identifiable community.

But strange as it may seem the Code intended to deal with them has remained in operation and is now applied to all low-class Hindus, who have no lock stock with the original *Sudras.* How this happened must be a matter of curiosity to all. My explanation is that the Sudras of the Indo-Aryan Society in course of time became so degraded as a consequence of the severity of the Brahmanical laws that they really came to occupy a very low state in public life. Two consequences followed from this. One consequence was a change in the connotation of the word Sudra. The word Sudra lost its original meaning of being the name of a particular community and became a general name for a low-class people without civilisation, without culture, without respect and without position.

The second consequence was that the widening of the meaning of the word Sudra brought in its train the widening of the application of the Code. It is in this way that the so-called *Sudras* of the present-day have become subject to the Code, though they are *not Sudras* in the original sense of the word. Be that as it may, the fact remains that the Code intended for the original culprits has come to be applied to the innocents.

If the Hindu law-givers had enough historical sense to realise that the original *Sudras* were different from the present-day low-class people, this tragedy—this massacre of the innocents—would have been avoided. The fact, however unfortunate it may be, is that the Code is applied to the present-day *Sudras* in the same rigorous manner in which it was applied to the original *Sudras.* How such a Code came into being cannot therefore be regarded as of mere antiquarian interest to the *Sudras* of today.

While it may be admitted that a study of the origin of the *Sudras* is welcome, some may question my competence to handle the theme. I have already been warned that while I may have a right to speak on Indian politics, religion and religious history of India are not my field and that I must not enter it. I do not know why my critics have thought it necessary to give me this warning. If it is an antidote to any extravagant claim made by me as a thinker or a writer, then it is unnecessary. For, I am ready to admit that I am not competent to speak even on Indian politics. If the warning is for the reason that I cannot claim mastery over the Sanskrit language, I admit this deficiency. But I do not see why it should disqualify me altogether from operating in this field. There is very little of literature in the Sanskrit language which is not available in English.

The want of knowledge of Sanskrit need not therefore be a bar to my handling a theme such as the present. For I venture to say that a study of the relevant literature, albeit in English translations, for 15 years ought to be enough to invest even a person endowed with such moderate intelligence like myself, with sufficient degree of competence for the task. As to the exact measure of my competence to speak on the subject, this book will furnish the best testimony. It may well turn out that this attempt of mine is only an illustration of the proverbial fool rushing in where the angels fear to tread. But I take refuge in the belief that even the fool has a duty to perform, namely, to do his bit if the angel has gone to sleep or is unwilling to proclaim the truth. This is my justification for entering the prohibited field.

4

Ambedkar Notes on History of India

More important for the history of India were the conquests of the Shakas and Yueh-chih, nomad tribes of Central Asia similar to the modern Turkomans The former are first heard of in the basin of the river Hi, and being dislodged by the advance of the Yueh-chih moved southwards reaching north-western India about 150 B. C. Here they founded many small principalities, the rulers of which appear to have admitted the suzerainty of the Parthians for sometime and to have borne the title of Satraps. It is clear that western India was parcelled out among foreign princes called Shakas, Yavanas, or Pallavas whose frontiers and mutual relations were constantly changing. The most important of these principalities was known as the Great Satrapy which included Surashtra (Kathiawar) with adjacent parts of the mainland lasted until about 395 A.D.

The Yueh-chih started westwards from the frontiers of China about 100 B. C. and, driving the Shakas before them, settled in Bacteria. Here Kadphises, the chief of one of their tribes, called the Kushans, succeeded in imposing his authority on the others who coalesced into one nation henceforth known by the tribal name. The chronology of the Kushan Empire is one of the vexed questions of Indian history and the dates given below are stated positively only because there is no space for adequate discussion and are given with some scepticism, that is desire for more knowledge founded on facts. Kadphises I (c. 15-45 A. D.) after consolidating his Empire led his armies southwards, conquering Kabul and perhaps Kashmir.

His successor Kadphises II (c. 45-78 A. D.) annexed the whole of north-western India, including northern Sindh, the Punjab and perhaps Benaras. There was a considerable trade between India and the Roman Empire at this period and an embassy was sent to Trojan, apparently by Kanishka (c. 78-123), the successor of Kadphises. This monarch played a part in the later history of Buddhism comparable with that of Ashoka in earlier ages He waged war with the Parthians and Chinese, and his Empire which had its capital at Peshawar included Afghanistan, Bacteria, Kashgar, Yarkhand, Khotan and Kashmir. These dominions, which perhaps extended as far as Gya in the east, were retained by his successors Huvishka (123-140 A. D.) and Vasudeva (140-178 A. D.) but after this period the Andhra and Kushan dynasties both collapsed as Indian powers, although Kushan kings continued to rule in Kabul. The reasons of their fall are unknown but may be connected with the rise of the Sassanids in Persia. For more than a century, the political history of India is a blank and little can be said except that the kingdom of Sliraṣtra continued to exist under a Shaka dynasty.

Light returns with the rise of the Gupta dynasty, which roughly marks the beginning of modern Hinduism and of a reaction against Buddhism. Though nothing is known of the fortunes of Pataliputra, the ancient imperial city of the Mauryas, during the first three centuries of our era, it continued to exist. In 320 a local Raja known as Candragupta I increased his dominions and celebrated his coronation by the institution of the Gupta era. His son Samudra Gupta continued his conquests and in the course of an extraordinary campaign, concluded about 340 A. D. appears to have received the submission of almost the whole peninsula. He made no attempt to retain all this territory but his effective authority was exercised in a wide district extending from the Hugli to the rivers Jumuna and Chambal in the west and from the Himalayas to the Narbuda. His son Candragupta II or Vikramaditya added to these possessions Malwa, Gujarat and Kathiawar and for more than half a century the Guptas ruled undisturbed over nearly all northern India except Rajputana and Sindh. Their capital was at first Pataliputra, but afterwards Kausambi and Ayodhya became royal residences.

The fall of the Guptas was brought about by another invasion of barbarians known as Huns, Ephthalites or White Huns and apparently a branch of the Huns who invaded Europe. This branch remained behind in Asia and occupied northern Persia. They invaded India first in 455, and were repulsed, but returned about 490 in greater force and overthrew the Guptas. Their kings Tormana and Mihiragula were masters of northern India till 540 and had their local capital at Sialkot in the Punjab, though their headquarters were rather in Barnyin and Baikh. The cruelties of Mihiragula provoked a coalition of Hindu princes. The Huns were driven to the north and about 565 A. D. their destruction was completed by the allied forces of the Persians and Turks. Though they founded no permanent states their invasion was important, for many of them together with kindered tribes such as the Gurjars (Gujars) remained behind when their political power broke up and, like the Shakas and Kushans before them, contributed to form the population of north-western India, especially the Rajput clans.

The defeat of the Huns was followed by another period of obscurity, but at the beginning of the seventh century Harsha (606-647 A. D.), a prince of Thanesar, founded after thirty five years of warfare, a state which though it did not outlast his own life, emulated for a time the dimensions and prosperity of the Gupta Empire. We gather from the account of the Chinese pilgrim Hsuan Chaung, who visited his court at Kannauj, that the kings of Bengal. Assam and Ujjain were his vassals but that the Panjab, Sindh and Kashmir were independent. Kalinga, to the south of Bengal was depopulated but Harsha was not able to subdue Pulakesin II, the Calukya king of the Deccan.

Let us now turn for a moment to the history of the south. It is even more obscure both in events and chronology than that of the north, but we must not think of the Dravidian countries as Uninhabited or barbarius. Even the classical writers of Europe had some knowledge of them. King Pandion (Pandya) sent a mission to Augustus in 20 B.C. Pliny speaks of Modura (Madura) and Ptolemy also mentions this town with about forty others. It is said that there was a temple dedicated to Augustus at Maziris, identified with Craganore. From an early period the extreme south of the peninsula was divided into three states known as the Pandya, Cera and Cola kingdoms

The first corresponded to the districts of Madura and Tinnevelly. Cera and Kerala lay on the west coast in the modern Travancore.

The Cola country included Tanjore, Trichinopoly, Madras, with the greater part of Mysore. From the sixth to the eighth century A. D. a fourth power was important, namely the Pallavas, who apparently came from the north of the Madras presidency. They had their capital at Canjeevaram and were generally at war with the three kingdoms. Their king, Narasimha-Varman (625-645 A. D.) ruled over part of the Deccan and most of the Cola country but after about 750 they declined, whereas the Colas grew stronger and Rajaraja (985-1018) whose dominions included the Madras Presidency and Mysore made them the paramount power in southern India, which position they retained until the thirteenth century.

As already mentioned, the Deccan was ruled by the Andhras from 220 B. C. to 236 A. D., but for the next three centuries nothing is known of its history until the rise of the Calukya dynasty at Vatapi (Badami) in Bijapur. Pulakesin II of this dynasty (608-642), a contemporary of Harsha, was for some time successful in creating a rival Empire which extended from Gujarat to Madras, and his power was so considerable that he exchanged embassies with Khusru II, King of Persia, as is depicted in the frescoes of Ajanta. But in 642 he was defeated and slain by the Palavas.

With the death of Pulakesin and Harsha begins what has been called the Rajput period, extending from about 650 to 1000 A. D. and characterized by the existence of numerous kingdoms ruled by dynasties nominally Hindu, but often descended from northern invaders or non-Hindu aboriginal tribes. Among them may be mentioned the following:

1. Kannauj or Panchala. This kingdom passed through troublous times after the death of Harsha but from about 840 to 910 A. D. under Bhoja (or Mihira) and his son, it became the principal power in northern India, extending from Bihar to Sindh. In the twelfth century it again became important under the Gaharwar dynasty.
2. Kannauj was often at war with the Palas of Bengal, a line of Buddhist kings which began about 730 A. D. Dharmapala (c. 800 A. D.) was sufficiently powerful to

depose the king of Kannauj. Subsequently the eastern portion of the Pala Kingdom separated itself under a rival dynasty known as the Senas.

3. The districts to the south of the Jumuna known as Jejak-abhukti (Bundelkhand) and Cedi (nearly equivalent to our Central Provinces) were governed by two dynasties known as Candels and Kalacuris. The former are thought to have been originally Gonds. They were great builders and constructed among other monuments the temples of Khajurao. Kirdvarman Chandel (1049-1100) greatly extended their territories. He was a patron of learning and the allegorical drama Prabodhacandrodaya was produced at his Court.
4. The Paramara (Pawar) dynasty of Malwa were-likewise celebrated as patrons of literature and kings Munja (974-995) and Bhoja (1018-1060) were authors as well as successful warriors.

Shaka Period

According to Vincent Smith, after first adopting A. D. 78 which appeared the most probable, finally chose 120 A. D. and we may agree that this date marks the beginning of the Shaka period inaugurated by Kanishka.

The order in which the chief Kushan kings followed doubtful. It is generally agreed that Kanishka cameaftre phises I (Kujula Kara Kadphises) and II (Vima Kadphises) former of these two, a Bactrinised Scythian, must, in Dr. Smith's view, have assumed power about 40 A. D. He seized Gandhara and the country of Taxila from Gondophares, the Parthian prince who, according to the apocryphal acts of the apostles, received St. Thomas. His son Vima (78-110) carved out a great empire for himself, embracing the Punjab and the whole western half of the Ganges basin.

There seems to have been an interval of about 10 years between Kadphises and Kanishka, the latter was the son of one Vajheshka and no relation of his predecessor, he seems to have been from Khotan, not Bacteria, and indeed he spent the summer at Kapisi in Paropan... and the winter at Purushapura (Peshawar) the axis of his empire was no longer in the (midst) of the Graeco-Iranian country. The empire of Kanishaka did not last long. Of his two sons, Vasishka and Havishka only the second survived him.

The power of the Kushans in the third century was reduced to Bacteria with Kabul and Gandhara, and they fell beneath the yoke of the Sassanids.

Kshatrappas or Satraps: This title, which is Iranian, is borne by two dynasties founded by the Shakas he had been driven from their country by the Yuch-chi invasion.

I. The first was established in Surashtra (Kathewar). One prince of this line Chasthana, seems to have held Malwa before the great days of the Kushans and to have become a vassal of Kanishka; he ruled over Ujjayini, which was the centre of the Indian civilisation.

II. The second line to which the name of Kshaharata is more particularly attached, was the hereditary foe of the Andhras; it ruled over Maharashtra, the country between modern Surat and Bombay. It was this latter Shaka state that was annihilated by the Satakarni and it was the former which arranged it, when Rodraman, the Satrap of Ujjayni conquered the Andhra King. The antagonism between the eastern & western states seems to have been accompanied by a difference of ideals. The Shakas, like all the Scythians of India or Serindia, such as the Thorkhans, retained from their foreign origin a sympathy for Buddhism, whereas the Andhras were keen supporters of Brahmanism.

The Guptas

The events of the third century are unknown to history and we have very, little information about the Kushan empire.

Day light returns in 318-19, when there arises in the old country of Magadh a new dynasty-Gupta.

The Guptas-Chandragupta II conquered the country of Malvas, Gujrathand Surashtra (Kathiwar) overthrowing the 1st great Satrap of the Shaka dynasty of Ujjain. As an extension of his territory westward he made Ayodhya and Kausambi his capitals instead of Pataliputra. About 155 (B.C.) he conquered the whole of the lower Indus and Kathewar, waged war in Rajputana, and Oudh but took Mathura (Muttra) on the

Jumuna, and even reached Pataliputra. He was severely defeated by Pushyamitra (?). Bactriana was at least in the north, a barrier between Parthia and India. India was therefore less exposed to attack from Parthia. Nevertheless, there was at least one Parthian ruler, Mithradates 1(171-136) who annexed the country of Taxila for a few years, about 138.

End of the Independence of Parthia and Bacteria

The event that put an end to the independence of Parthia and Bacteria was a new invasion, resulting from a movement of tribes, which had taken place far away from India in the Mongolian steppes.

About 170 (B.C.) a horde of nomadic Scythians, the Yuch-chi or Tokharians, being driven from Gobi, the present Kansu, by the Hiang-nu or Huns, started on a wild migration which upset the whole balance of Asia.

They fell on the Shakas, who were Iranianised Scythians dwelling north of the Persion empire and settled in their grazing grounds north of the Jazartes. The expelled Shakas fell on Parthia and Bactriana, obliterating the last vestiges of Greek rule, between 140 and 120 (B. C.) Then the Tokharians, being defeated in their turns by the Wu-Sun tribe, established themselves on the Oxus, and after that took all the country of the Shakas in eastern Iran at the entrance to India. That entrance was found in the first century after Christ.

The conquest of India was the work of the Kushans (Kushana), a dynasty which united the Yue-Chi tribes and established their dominion both over their own kinsfolk the Shakas of Parthia and over peoples of the Punjab. The accession of the principal King of this line, Kanishka, was placed at uncertain dates between 57 B. C. and A. D. 200.

Pushyamitra—a mayor of the Palace as Sybrani Livi called him.

The Selected Empire ruled by Antiochos III (261-246 B.C.) and lost two provinces Parthia and Bactriana which emancipated themselves simultaneously. The Parthians whom the Indians called Pahalvas, were related to the nomads of the Turkoman steppes and occupied the country south-east of the Caspian. The Bacterins bordered on the Parthians on the north-east and were settled between the Hindu Kush and the Oxus; the number and wealth of their towns were legendary. These two peoples seem to have taken advantage of the difficulties of Antiochos and his successors, Seleucos II (246-226 B.C.) and III (226-223 B.C.) in the west to break away.

The Parthian revolt was a natural movement, led by Arsaces, the founder of a dynasty which was to rule Persia for nearly 500 years. The Bactrian rising was brought about by the ambition of a Greek satrap. Diodotos, represents an outbreak of Hellennism in the heart of Asia. There is no doubt that the formation of these enterprising nations on the Indo-Iranian border helped to shake the empire of Ashoka in the time of his successors.

The Punjab, once a Persian satrapy and then a province of Alexander, was to find itself still more exposed to attack, now that smaller but turbulent states had arisen at its doors. After Diodotos I & II, the King of Bacteria was Euthidemes, who went to war with Antioches the Great of Syria. Peace was concluded with the recognition of Bactrian independence about 208. But during hostilities Syrian troops had crossed the Hindu Kush and entering the Kabul valley had severely despoiled the ruler Subhagasena. Demetrius, the son of Enthidemos, increased his dominion not only in the present Afghanistan but in India proper, and bore the title of King of the Indians (200-190). Between 190 and 180 there were Greek adventurers reigning at Taxila, named Paleon & Agathocles. From 160 to 140 roughly, Kabul and the Punjab were held by a pure Greek, Milinda or Minander, who left a name in the history of Buddhism.

Huns

In the last years of Kumargupta new Iranian peoples assailed the empire, but they were kept back from the frontiers. Under Skandagupta, the first wave of formidable migration came down upon the same frontiers. This consisted of nomad Mongoloids to whom India afterwards gave the genuine name of Huna, under which we recognised the Huns who invited Europe.

Those who reached India after the middle of the fifth century were white Huns or Ephthalites, who in type were closer to the

Turks than to the hideous followers of Attila. After a halt in the valley of the Oxus they took possession of Persia and Kabul. Skandagupta had driven them off for a few years (455 A. D.) but after they had slain Firoz the Sassanid in 484, no Indian state could stop them. One of them, named Toramana, established himself among the Malavas in 500 and his son Mihirgula set up his capital at Sakol (Sialkot) in the Punjab. A native prince Yeshodharman shook off the yoke of Mihirgula. The expulsion of the Huns was not quite complete everywhere. A great many resided in the basin of the Indus.

At the beginning of the 7th century a power arose from the chaos in the small principality of Sthanvisvara (Thaneshwar, near Delhi). Here a courageous Raja Prabhakar Vardhan organised a kingdom, which showed its mettle against the Gurjars, the Malwas and other neighbouring princes. Shortly after his death in 604 or 605 his eldest son was murdered by the orders of the king of Gauda in Bengal. The power fell to his younger brother Harsha.

Manu and the Sudras

The reader is now aware that in the Scheme of Manu there were two principal social divisions: those outside the Chaturvarna and those inside the Chaturvarna. The reader also knows that the present day Untouchables are the counterpart of those outside the Chaturvarna and that those inside the Chaturvarna were contrasted with those outside. They were a composite body made up of four different classes, the Brahmins, the Kshatriyas, the Vaishyas and the Sudras. The Hindu social system is not only a system in which the idea of classes is more dominant than the idea of community but it is a system which is based on inequality between classes and therefore between individuals.

To put it concretely, the classes *i.e.* the Brahmins, Kshatriyas, Vaishyas, Sudras and Antyajas (Untouchables) are not horizontal, all on the same level. They are vertical *i.e.* one above the other. No Hindu will controvert this statement. Every Indian knows it. If there is any person who would have any doubt about it he can only be a foreigner. But any doubt which a foreigner might have will be dissolved if he is referred to the law of Manu who is the chief architect of the Hindu society and whose law has formed the foundations on which it is built. For his benefit I reproduce such texts from the Manusmriti as go to prove that Hindu society is based on the principle of inequality.

It might be argued that the inequality prescribed by Manu in his Smriti is after all of historical importance. It is past history and cannot be supposed to have any bearing on the present conduct of the Hindu. I am sure nothing can be greater error than this. Manu is not a matter of the past. It is even more than a past of the present. It is a 'living past' and therefore as really present as any present can be.

That the inequality laid down by Manu was the law of the land under the pre-British days may not be known to many foreigners. Only a few instances will show that such was the case. Under the rule of the Marathas and the Peshwas the Untouchables were not allowed within the gates of Poona city, the capital of the Peshwas between 3 p.m. and 9 a.m. because, before nine and after three, their bodies cast too long a shadow; and whenever their shadow fell upon a Brahmin it polluted him, so that he dare not taste food or water until he had bathed and washed the impurity away. So also no Untouchable was allowed to live in a walled town; cattle and dogs could freely enter but not the Untouchables

Under the rule of the Marathas and the Peshwas the Untouchables might not spit on the ground lest a Hindu should be polluted by touching it with his foot, but had to hang an earthen pot round his neck to hold his spittle. He was made to drag a thorny branch of a tree with him to brush out his footsteps and when a Brahman came by, had to lie at a distance on his face lest his shadow might fall on the Brahman

In Maharashtra an Untouchable was required to wear a black thread either in his neck or on his wrist for the purpose of ready identification.

In Gujarat the Untouchables were compelled to wear a horn as their distinguishing mark.

In the Punjab a sweeper was required while walking through streets in towns to carry a broom in his hand or under his armpit as a mark of his being a scavenger.

In Bombay the Untouchables were not permitted to wear clean or untorn clothes. In fact the shopkeepers took the precaution to see that before cloth was sold to the Untouchable it was torn & soiled.

In Malabar the Untouchables were not allowed to build houses above one storey in height and not allowed to cremate their dead.

In Malabar the Untouchables were not permitted to carry umbrellas, to wear shoes or golden ornaments, to milk cows or even to use the ordinary language of the country.

In South India Untouchables were expressly forbidden to cover the upper part of their body above the waist and in the case of women of the Untouchables they were compelled to go with the upper part of their bodies quite bare.

In the Bombay Presidency so high a caste as that of Sonars (goldsmiths) was forbidden to wear their Dhoties with folds and prohibited to use *Namaskar* as the word of salutation#.

The following letter will be interesting to the reader as it throws a flood of light as to whether the Dhamia prescribed by Manu was or was not the law of the land-

"To

Damulsett Trimbucksett

Head of the Caste of Goldsmiths.

"The Hon'ble the President in Council having thought proper to prohibit the Caste of Goldsmiths from making use of the form of salutation termed Namaskar, you are hereby pre-emptorily enjoined to make known this order and resolution to the whole caste and to take care that the same be strictly observed.

By order

Secretary to Government

sig. W. Page

Bombay

9th August 1779

Resolution of Government

Dated 28th July 1779.

"Frequent disputes having arisen for some time between the Brahmins and Goldsmiths respecting a mode of salutation termed "Namaskar" made use of by the latter, and which the Brahmins allege they have no right to perform, and that the exercise of such ceremony by the Goldsmiths is a great breach and profanation of the rights of the Gentoo (Hindu] Religion, and repeated complaints having been made to us by the Brahmins, and the Peishwa also having several times written to the President, requesting the use of the Namaskar might be prohibited to the Goldsmiths-Resolved as it i« necessary. This matter should be decided by us in order that the dispute between the two castes may be put an end to, and the Brahmins appear to have reason for their complaint, that the Goldsmiths be forbidden the use of the Namaskar, and this being a matter wherein the Company's interest is not concerned, our Resolution may be put on the footing of a compliment to the Peshwa whom the President is desired to make acquainted with our determination."

Under the Maratha rule any one other than a Brahmin uttering a Veda Mantra was liable to have his tongue cut off and as a matter of fact the tongues of several Sonars (goldsmiths) were actually cut off by the order of the Peshwa for their daring to utter the Vedas contrary to law. All over India Brahmin was exempt from capital punishment. He could not be hanged even if he committed murder.

Under the Peshwas distinction was observed in the punishment of the criminals according to the caste. Hard labour and death were punishments mostly visited on the Untouchables.

Under the Peshwas Brahmin clerks had the privilege of their goods being exempted from certain duties and their imported corn being carried to them without any ferry charges; and Brahmin landlords had their lands assessed at distinctly lower rates than those levied from other classes. In Bengal the amount of rent for land varied with the caste of the occupant and if the tenant was an Untouchable he had to pay the highest rent.

These facts will show that Manu though born some time before B. C. or sometime after A. D. is not dead and while the

Hindu Kings reigned, justice between Hindu and Hindu, touchable and untouchable was rendered according to the Law of Manu and that law was avowedly based on inequality.

This is the Dharma laid down by Manu. It is called Manav Dharma *i.e.* Dharma which by its inherent goodness can be applied to all men in all times and in all places. Whether the fact that it has not had any force outside India is a blessing or a curse I do not stop to inquire. It is important to note that this Manav Dharma is based upon the theory that the Brahman is to have all the privileges and the Sudra is not to have even the rights of a human being, that the Brahman is to be above everybody in all things merely by reason of his high birth and the Sudra is to be below everybody and is to have none of the things no matter how great may be his worth.

Nothing can show the shamelessness and absurdity of this Manava Dharma better than turning it upside down. I know of no better attempt in this behalf than that of Dr. R. P. Pranjape a great Educationist, Politician and Social reformer and I make no apology for reproducing it in full.

Peep Into the Future

This piece Was written against the Non-Brahmin Parties which were then in power in the Bombay and Madras Presidency and in the Central Provinces. The Non-Brahmin parties were founded with the express object of not allowing a single community to have a monopoly in State Service. The Brahmins have a more or less complete monopoly in the State services in all provinces in India and in all departments of State. The Non-Brahmin parties had therefore laid down the principle, known as the principle of communal ratio, that given minimum qualifications candidates belonging to non-Brahmin communities should be given preference over Brahmin candidates when making appointments in the public services. In my view there was nothing wrong in this principle. It was undoubtedly wrong that the administration of the country should be in the hands of a single community however clever such a community might be.

The Non-Brahmin Party held the view that good Government was better than efficient Government was not a principle to be confined only to the composition of the Legislature & the Executive. But that it must also be made applicable to the field of administration. It was through administration that the State came directly in contact with the masses. No administration could do any good unless it was sympathetic. No administration could be sympathetic if it was manned by the Brahmins alone.

How can the Brahmin who holds himself superior to the masses, despises the rest as low caste and Sudras, is opposed to their aspiration, is instinctively led to be partial to his community and being uninterested in the masses is open to corruption be a good administrator? He is as much an alien to the Indian masses as any foreigner can be. As against this the Brahmins have been taking their stand on efficiency pure & simple. They know that this is the only card they can play successfully by reason of their advanced position in education. But they forget that if efficiency was the only criterion then in all probability there would be very little chance for them to monopolise State service in the way and to the extent they have done. For if efficiency was made the only criterion there would be nothing wrong in employing Englishmen, Frenchmen, Germans & Turks instead of the Brahmins of India. Be that as it may, the Non-Brahmin Parties refused to make a fetish to efficiency and insisted that there must be introduced the principle of communal ratio in the public services in order to introduce into the administration an admixture of all castes & creeds and thereby make it a good administration.

In carrying out this principle the Non-Brahmin Parties in their eagerness to cleanse the administration of Brahmindom while they were in power, did often forget the principle that in redressing the balance between the Brahmins and non-Brahmins in the public services they were limited by the rule of minimum efficiency. But that does not mean that the principle they adopted for their guidance was not commendable in the interests of the masses.

This policy no doubt set the teeth of many Brahmins on edge. They were vehement in their anger. This piece by Dr. Paranjpe is the finest satire on the policy of the non-Brahmin Party. It caricatures the principle of the non-Brahman party in a manner which is inimitable and at the time when it came out, I know many non-Brahmin leaders were not only furious but also speechless. My complaint against Dr. Paranjpe is that he did not see the humour of it. The non-Brahmin Party was doing nothing new. It was merely turning Manusmriti upside down.

It was turning the tables. It was putting the Brahmin in the position in which Manu had placed the Sudra. Did not Manu give privileges to Brahmin merely because he was a Brahmin? Did not Manu deny any right to the Sudra even though he deserved it? Can there be much complaint if now the Sudra is given some privileges because he is a Sudra? It may sound absurd but the rule is not without precedent and that precedent is the Manusmriti itself. And who can throw stones at the non-Brahmin Party? The Brahmins may if they are without sin. But can the authors and worshippers, upholders of Manusmriti claim that they are without sin? Dr. Paranjpe's piece is the finest condemnation of the inquiry that underlies this Manav Dharma. It shows as nothing else does what a Brahmin feels when he is placed in the position of a Sudra.

Inequality is not confined to Hindus. It prevailed elsewhere also and was responsible for dividing society into higher and lower free and servile classes. (Left incomplete in Ms.-ed.)

Preservation of Social Order

The second technique devised for the maintenance and preservation of the Established Order is quite different from the first. Really speaking it is this, which constitutes a special feature of the Hindu Social Order.

In the matter of the preservation of the Social Order from violent attack it is necessary to bear in mind three considerations. The outbreak of a Revolution is conditioned by three factors.

(1) The existence of a sense of wrong,

(2) Capacity to know that one is suffering from a wrong and

(3) Availability of arms.

The second consideration is that there are two ways of dealing with a rebellion. One is to prevent a rebellion from occurring and the other is to suppress it after it has broken out. The third consideration is that whether the prevention of rebellion would be feasible or whether the suppression of rebellion would be the only method open, would depend upon the rules which govern the three prerequisites of rebellion.

Where the Social Order denies opportunity to rise, denies right to education, and denies right to use arms it is in a position to prevent rebellion against the Social Order. Where on the other hand a Social Order allows opportunity to rise, allows right to education, and permits the use of arms it cannot prevent rebellion by those who suffer wrongs. Its only remedy to preserve the Social Order by suppressing rebellion by the use of force and violence. The Hindu Social Order has adopted the first method. It has fixed the social status of the lower orders for all generations to come. Their economic status is also fixed. There being no disparity between the two there is no possibility of a grievance growing up. It has denied education to the lower orders.

The result is that no one is conscious that his low condition is a ground for grievance. If there is any consciousness it is that no one is responsible for the low condition. It is the result of fate. Assuming there is a grievance, assuming there is consciousness of grievance there cannot be a rebellion by the lower orders against the Hindu Social Order' because the Hindu Social order denies the masses the right to use arms.... Social Orders such as use of follow the opposite course. They allow equal opportunity to all. They allow freedom to acquire knowledge, they allow the right to bear arms and take upon themselves the odium of suppressing rebellious force & violence. To deny freedom of opportunity, to deny freedom to acquire knowledge, to deny the right to arms is a most cruel wrong. Its result to and man. The Hindu Social Order is not ashamed to this. It has however achieved two things. It has found the most effective even though it be the most shameless method of preserving the established Order. Secondly notwithstanding the use of most inhuman means of killing manliness, it has given to the Hindus the reputation of being a very humane people.

Another special feature of the Hindu Social Order relates to the technique devised for its preservation.

The technique is twofold.

The first technique is to place the responsibility of upholding and maintaining the social order upon the shoulders of the King. Manu does this in quite express terms.

VIII. 410. "The king should order each man of the mercantile class to practice trade, or money-lending, or agriculture and attendance on cattle; and each man of the servile class to act in the service of the twice-born."

VIII. 418. "With vigilant care should the King exert himself in compelling merchants and mechanics to perform their respective duties; for, when such men swerve from their duty, they throw this world into confusion."

Manu does not stop with the mere enunciation of the duty of the King in this behalf. He to ensure that the King shall at all times perform his duty to maintain and preserve the Established Order. Manu therefore makes two further provisions. One provision is to make the failure of the King to maintain the Established Order an offence for which the King become liable for prosecution and punishment like a common man. This would be clear from the following citations from Manu:

VIII. 335. "Neither a father, nor a preceptor, nor a friend, nor a mother, nor a wife, nor a son, nor a domestic priest must be left unpunished by the king, if they adhere not with firmness to their duty."

VIII. 336. "Where another man of lower birth would be fined one pana, the king shall be fined a thousand, and he shall give the fine to the priests, or cast it into the river, this is a sacred rule."

The other provision made by Manu against a King who is either negligent or opposed to the Established Order is to invest the three classes. Brahmins, Kshatriya and Vaishya with a right to rise in armed rebellion against the King.

VIII. 348. "The twice-born may take arms, when their duty is obstructed by force; and when, in some evil time, a disaster has befallen the twice-born classes."

With the Hindus

It is impossible to believe that Hindus will ever be able to absorb the Untouchables in their society. Their Caste System and the Religion completely negative any hope being entertained in this behalf.

Yet there are incorrigible optimists more among the Hindus than among the Untouchables, who believe in the possibility of the Hindus assimilating the Untouchables. Whether these incorrigible optimists are honest or dishonest in their opinion is a question which cannot be overlooked. Within what time this assimilation will take place, they are unable to define. Assuming that the optimists are honest, there can be no question that this process of assimilation is going to be a long drawn process extending over many centuries.

In the meantime the Untouchables will have to live under the Social and political sway of the Hindus, and continue to suffer all the tyrannies and oppressions to which they have been subjected in the past. Obviously no sane man will think of leaving them to the will and the pleasure of the Hindus in the hope that some day in the unpredictable future they will be assimilated by the Hindus. Long or short, there will be a period of transition and some provision must be made against the tyranny and oppression by the Hindus. What provisions should be made in this behalf? If the question is left to the Untouchables they will ask for two provisions being made: one for Constitutional Safeguards and two for Separate Settlements.

The nature of Constitutional Safeguards for the protection of the Untouchables have been defined by the All-India Scheduled Caste Federation, a political organisation of the Untouchables of India in the form of resolutions.

The Hindus are very reluctant to allow the Untouchables these safeguards. The objection is general. There is also objection to particular safeguards. The general objection that the Untouchables are not a minority and therefore they are not entitled to safeguards which may be allowed to other minorities. The argument proceeds that the basis of a community to be called a minority is Religion if one is entitled to be recognised as a minority. The Untouchables are not separate from the Hindus in the matter of the religion. Consequently they are not a minority. That this definition of a minority is childish will be obvious to all those who have studied the question.

Frustration

The Untouchables are the weariest, most loathed and the most miserable people that history can witness. They area spent and sacrificed people. To use the language of Shelley they are—"pale for weariness of climbing heaven, and gazing on earth, wandering companionless Among the stars that have a different birth"

To put it in simple language the Untouchables have been completely overtaken by a sense of utter frustration. As Mathew Arnold says "life consists in the effort to affirm one's own essence; meaning by this, to develop one's own existence fully and freely, to have ample light and air, to be neither (.......) nor overshadowed. Failure to affirm ones own essence is simply another name for frustration. Its non fulfilment of one's efforts to do the best, the withering of one's faculties, the stunting of one's personality."

Many people suffer such frustrations in their history. But they soon recover from the blight and rise to glory again with new vibrations. The case of the Untouchables stands on a different footing. Their frustration is frustration for ever. It is unrelieved by space or time. In this respect the story of the Untouchables stands in strange contrast with that of the Jews.

Their captivity in Egypt was the first calamity that visited the Jewish people. As the Bible says.

Ultimately Pharaoh yielded. The Jewish people escaped captivity and went to Cannan and settled there in the land flowing with milk and honey.

The second calamity which overtook the Jews was the Babylonian Captivity.

We can now explain why the Untouchables have suffered frustration. They have no plus condition of body and mind. They have nothing in their dull drab deadening past for a hope of a rise in the future to feed upon. This is due to no fault of theirs. The frustration which is their fate is the result of the unpropitious social environment born out of the Hindu Social Order which is so deadly inimical to their progress.

The Covenant with God may be interpreted to mean in the language of Emerson a plus condition of mind and body. As

Emerson has said "Success is constitutional-depends, on a plus condition of mind and body—on power of work—on courage. Success goes invariably with a certain plus or positive power: An ounce of Power must balance an ounce of weight."

If the Jews rose after their first captivity, it was primarily because of their plus condition of mind and body. This plus condition of mind and body can arise from two sources. It can arise from reliance on God. God, if nothing else is at least a source of power and in emergency man needs mental power, the plus condition of mind and body which is necessary for success. There is therefore nothing wrong in the suggestion that the Jews succeeded because of their Covenant of God if it is interpreted in the right way.

This plus condition of body and mind is also the result of Social Environment, if the Environment is propitious. In a society where there is exemption from restraint, a secured release from obstruction, in a society where every man is entitled not only to the means of being, but also of well-being, where no man is forced to labour so that another may abound in luxuries, where no man is deprived of his right to cultivate his faculties and powers so that there may be no competition with the favoured, where there is emphasis of reward by mento, where there is goodwill towards all.

The Problem of Political Suppression

The introduction of the principle of political liberty in India has been very tardy & gradual. It began in the year 1892 when the principle of popular representation in the Constitution of the Legislatures was introduced. It was expanded in 1909. There were two defects in the popular representation as it stood in 1909. The first defect was the franchise was very high. It was so high that a large mass of people were excluded. Those to whom it reached were the aristocracy of the Hindus and the Muslims. The second defect was that the scheme of popular representation was confined to the Legislature. It did not extend to the Executive.

The Executive continued to be independent. The Legislature could neither make or unmake the Executive. The next was taken in 1919. Curiously enough, in the scheme of 1919 the principle of popular representation was applied to the Executive without applying it in commensurate degree to the Legislature. This happened because the political movement in India was led largely by the higher classes. They have always been more anxious for Executive power than for extension of franchise. It is natural. For they stood to gain by executive power. While those who would gain by franchise were the masses.

The higher classes having the ear of the British authorities pressed for executive power and succeeded getting it without the extension of franchise. The franchise was no doubt extended much beyond the bounds fixed in 1909. But it did not touch the Untouchables. Indeed they are so poor that nothing except adult franchise would bring the Untouchables on the electoral roll.

The Government of India was very much perturbed. They could do very little. But they did express their anxiety about placing the Untouchables under the political domination of the high caste Hindus without giving the Untouchables the right to vote in the election. In their despatch of 19th March 1919 the Government of India observed— [Quote] The situation was altered in 1935 under the scheme proposed by the British Government under what is called the Communal Award.

(i) The Untouchables were to have a differential franchise so as to enfranchise about ten per cent of their population.

(ii) The Untouchables were not only to have a differential franchise, they were to have certain number of seats reserved for them in the Provincial and Central Legislatures.

(iii) The seats reserved for them were to be filled by separate Electorates formed exclusively of voters belonging to the Untouchable Community.

(iv) In addition to having a vote in the Separate Electorates the Untouchables were to have a second or additional vote in the general election for seats open to Hindus other than the Untouchables.

Mr. Gandhi who had been objecting to separate representation of the Untouchables raised a protest against the proposal of the British Government and threatened to fast unto death if these concessions were not withdrawn. Mr. Gandhi's objection was mainly to Separate Electorates and as the British Government refused to withdraw their proposals unless there was an agreement between the Untouchables and the Hindus. There upon Mr. Gandhi started his fast. Eventually an agreement was arrived at between the Hindus and the Untouchables in September 1932. That agreement is known as the Poona Pact. Its terms are reproduced below:

Poona Pact

(1) There shall be seats reserved for the Depressed Classes out of the general electorate seats in the Provincial Legislatures as follows:

Madras	30
Bombay with Sindh	15
Punjab	8
Bihar and Orissa	18
Central Provinces	20
Assam	7
United Provinces	20
Total	148

These figures are based on the total strength of the Provincial Councils, announced in the Prime Minister's decision.

(2) Election to these seats shall be by joint electorates subject, however, to the following procedure:
All the members of the Depressed Classes registered in the fengral electoral roll in a constituency will form an electoral college, which will elect a panel of four candidates belonging to the Depressed Classes for each of such reserved seats, by the method of the single vote; the four persons getting the highest number of votes in such primary election, shall be candidates for election by the general electorate.

(3) Representation of the Depressed Classes in the Central Legislature shall likewise be on the principal of joint

electorates and reserved seats by the method of primary election in the manner provided for in Clause two above, for their representation in the Provincial Legislatures.

(4) In the Central Legislatures, eighteen per cent of the seats allotted to the general electorate for British India in the said legislature shall be reserved for the Depressed Classes.

(5) The system of primary election to a panel of candidates for election to the Central and Provincial Legislatures, as here in before mentioned, shall come to an end after the first ten years, unless terminated sooner by mutual agreement under the provision of Clause six below.

(6) The system of representation of the Depressed Classes by reserved seats in the Provincial and Central Legislature as provided *for* in Clauses I and 4 shall continue until determined by mutual agreement between the communities concerned in the settlement.

(7) Franchise for the Central and Provincial Legislatures for the Depressed Classes shall be as indicated in the Lothian Committee Report.

(8) There shall be no disabilities attaching to any one on the ground of his being a member of the Depressed Classes in regard to any elections to local bodies or appointment to the Public Services. Every endeavour shall be made to secure fair representation of the Depressed Classes in these respects, subject to such educational qualifications as may be laid down for appointment to the Public Services.

(9) In every province out of the educational grant, an adequate sum shall be earmarked for providing educational facilities to the members of the Depressed Classes.

This pact forms the charter of the political liberty of the Untouchables. The first election.......

Which is Worse?—Slavery or Untouchability?

Slavery in India

Among the claims made by the Hindus for asserting their

superiority over other nations the following two are mentioned. One is that there was no slavery in India among the Hindus and the other is that Untouchability is infinitely less harmful than slavery.

The first statement is of course untrue. Slavery is a very ancient institution of the Hindus. It is recognised by Manu, the law giver and has been elaborated and systematised by the other Smriti writers who followed Manu. Slavery among the Hindus was never merely ancient institution, which functioned, only in some hoary past. It was an institution which continued throughout all Indian history down to the year 1843 and, if it had not been abolished by the British Government bylaw in that year, it might have continued even today. While slavery lasted it applied to both the touchables as well as the untouchables.

The untouchables by reason of their poverty became subject to slavery of tener than did the touchables. So that up to 1843 the untouchables in India had to undergo the misfortune of being held in double bondage-the bondage of slavery and the bondage of untouchability. The lighter of the bonds has been cut and the untouchable is made free from it. But because the untouchables of today are not seen wearing the chains of slavery on them, it is not to be supposed that they never did. To do so would be to tear off whole pages of history.

The first claim is not so widely made. But the second is. So great a social reformer and so great a friend of the untouchables as Lala Lajpat Rai in replying to the indictment of the Hindu Society by Miss Mayo insisted that untouchability as an evil was nothing as compared with slavery and he fortified his conclusion by a comparison of the Negro in America with the untouchables in India and showed that his conclusion was true. Coming as it does from Lala Lajpat Rai the matter needs to be more closely examined.

Is untouchability less harmful than slavery? Was slavery less human than untouchability? Did slavery hamper the growth more than untouchability does? Apart from the controversy raised by Lala Lajpat Rai, the questions are important and their discussions will be both interesting and instructive. To understand this difference it is necessary to begin by stating the precise meaning of the term slavery. This is imperative because the term slavery is also used in a metaphorical sense to cover social relationship which is kindered to slavery but which is not slavery.

Because the wife was entirely in the power of the husband, because he sometimes ill-used her and killed her, because the husband exchanged or lent his wife and because he made her work for him, the wife was sometimes spoken of as a slave. Another illustration of the metaphorical use of the term is its application to I serfs. Because a serf worked on fixed days, performed fixed I services, paid fixed sums to the lord and was fixed to the land, he was spoken of as a slave. These are instances of curtailment of I freedom, and inasmuch as they are akin to slavery because slavery also involves loss of freedom. But this is not the sense in which the word is used in law, and to avoid arguing at cross purpose, it would be better to base the comparison on the legal meaning of the word slavery.

In layman's language, a person is said to be slave when he is the property of another. This definition is perhaps too terse for the lay reader. He may not understand the full import of it without further explanation, property means something, a term which is used to denote a bundle of rights which a person has over something which is his property, such as the right to possess, to use, to claim the benefit of, to transfer by way of sale, mortgage or lease and destroy. Ownership therefore means complete dominion over property. To put it concretely, when it is said that the slave is the property of the master, what it means is that the master can make the slave work against his will, take the benefit of whatever the slave produces without the consent of the slave.

The master can lease out, sell or mortgage his slave without consulting the wishes of the slave and the master can even kill him in the strictest legal connotation of the term. In the eye of the law the slave is just a material object with which his master may deal in any way he likes.

In the light of this legal definition, slavery does appear to be worse than untouchability. A slave can be sold, mortgaged or leased; an untouchable cannot be sold, mortgaged or leased. A slave can be killed by the master without being held guilty for murder; an untouchable cannot be. Whoever causes his death will be liable for murder. In fact, the slave could not be killed with impunity, the law did recognise his death as being culpable homicide as it did in the case of the death of a freeman. But taking the position of the slave as prescribed by laws the difference between the condition of the slave and the untouchable is undoubtedly clear-that the slave was worse off than the untouchable.

There is however another way of defining a slave which is equally legal and precise although it is not the usual way. This other way of defining a slave is this; A slave is a human being who is not a person in the eye of the law. This way of defining a slave may perhaps puzzle some. It may therefore be necessary to state that in the eye of the law the term person is identical with the term human being. In law, there may be human beings whom the law does not regard as persons. Contrariwise there are in law persons who are not human brings.

This curious result arises of the meaning which the law attaches to the word person. For the purposes of law a person is defined as an entity, human or non-human, in whom the law recognised a capacity for acquiring rights and bearing duties, A slave is not a person in the eye of the law although he is a human being. An idol is a person in the eye of the law although an idol is an inanimate object. The reason for this difference will be obvious. A slave is not a person although he is a human being, because the law does not regard him as an entity endowed with the capacity for rights and duties. Concisely an idol is a person though not a human being because the law does-whether wisely or not is another question-recognise the capacity for rights and duties. To be recognised as a person is of course a very important fact fraught with tremendous consequences. Whether one is entitled to rights and liberties upon this issue, the rights which flow from this recognition as person are not only as life but are as vital as life.

They include right over material things, their acquisition, their enjoyment and their disposal—called right to property. There are others far more important than these rights over material things. Firstly, there is the right in respect of one's own person—a right not to be killed, maimed or injured without due process of law called a right to life, a right not to be imprisoned save in due process of law-called right to liberty. Secondly, there is a right to reputation-a right not to be ridiculed or lowered in the estimation of fellow men, the right to his good name *i.e.* the right to the respect so far as it is well founded which others feel for him shall not be diminished. Thirdly, there is the right to the free exercise of powers and liberties

Every person is entitled without molestation to perform all lawful acts and to enjoy all the privileges which attach to him as a person. The most specific right of this kind is to be the unmolested pursuit of the occupation by which a man chooses to gain his livelihood. Under the same head falls the right of every person to the free use of the public highways, of navigable rivers and all public utilities. It also includes the right of every person that the machinery of the law, which is established for the protection of all persons shall not be maliciously set in motion to his detriment. Thirdly, there is the right of immunity from damage by fraud or coercion-it is a right not to be induced by fraud to assent to a transaction which causes damage, and not to be coerced into acting contrary to one's desire by force.

Fourthly, the rights of a person are those which are collectively called Family Rights. These family rights may be distinguished as 'marital', 'parental', 'tutelary', and 'dominical'. The marital right, the right of a husband as against the world, is that no other man shall, by force or persuasion, deprive him of his wife's society, still less be criminally intimate with her. An analogous right might conceivably be recognised as being vested in the wife and is recognised in parts of America. The parental right extends to the custody and control of children, to the produce of their labour till they arrive at the age of discretion without interference. The tutelary right is the right of the parent to act as the guardian not for the benefit of the guardian but for that of the ward......... whose want of understanding he supplements and whose affairs he manages. The dominical right is the right to use labour of the ward. The right is infringed by killing, by injuring so as to make him less valuable or by enticing him away.

Not being a person, a slave had, so far as law is concerned, none of these rights. The untouchable is a person in the eye of the law. It cannot therefore be said that he has none of the rights which the law gives to a 'person'. He has the right to property, to life, liberty, reputation, family and to the free exercise of his liberties and his powers. Define the slave as one may, either as a piece of property or as one who is not a person, *it* appears that the slave was worse off than the untouchable.

This is so if we consider only the *de jure* position of the slave. Let us consider what was the *defacto* position of the slave in the Roman Empire and in the United States. I take the following extracts from Mr. Barrow:

> *"Hitherto, it is the repulsive side of household slavery that has been sketched. There is also another aspect. The literature reveals the vast household as normal. It is, of course, the exception. Large slave staffs undoubtedly existed, and they are generally to be found in Rome. In Italy and the Provinces there was less need of display; many of the staff of the Villa were engaged in productive work connected with land and its produce. The old-fashioned relationship between foreman and slave remained there; the slave was often a fellow worker. The kindliness of Pliny towards his staff is well-known. It is in no spirit of self-righteousness and in no wish to appear in a favourable light in the eyes of the future generations which he hoped would read his letters that he tells of his distress at the illness and death of his slaves. The household (of Pliny) is the salves' republic. Pliny's account of his treatment of his slaves is sometimes regarded as so much in advance of general or even occasional practice as to be valueless as evidence. There is no reason for this attitude."*

From reasons both of display and genuine literary interest, the rich families attached to their households, slaves trained in literature and art. Calvisices Sabinus is said by Seneca to have had eleven slaves taught to recite Homer, Hesioid, and nine lyric poets by heart. ' Book cases would be cheaper ', said a rude friend. ' No, what the household knows the master knows ' was the answer. But, apart from such abuses, educated slaves must have been a necessity in the absence of printing;..... The busy lawyer, the dilettante poet, the philosopher and educated gentlemen of literary tastes had need of copyists and readers and secretaries. Such men were naturally linguistic also; a librarius who dies at the age of twenty boasts that he was ' *literatus Graecis at Latinis*'. *Amanuensis* were common enough; librarians are to be found in public and private libraries.....Shorthand writing was in common use under the Empire, and slave Notary were regularly employed....

Many freemen, rhetoricians and grammarians are collected by Snetonius in a special treatise. Verrius Flaccus was tutor to Austus's grandsons, and at death was publicly honoured by a statue. Scribonius Aphrodisius was the slave and disciple of Orbilius and was afterwards freed by Scribenia. Hyginus was librarian of the Palatine Library, in which office he was followed by Jullius Modestus, his own freedman. We hear of freedmen historians of a slave philosopher who was encouraged to argue with his master's, friends' slaves and freed architects. Freemen as doctors occur frequently in the inscriptions, some of them specialists; they had been trained in big households as slaves, as is shown by one or two examples; after Manumission they rose to eminence and became notorious for their high fees."

"The tastes of some section of society demanded that dancers, singers, musicians, montebanks, variety artists, athletic trainers and messeiurs should be forthcoming. All these are to be found in slavery, often trained by teachers who had acquired some reputation."

"The age of Augustus was the beginning of a period of commercial and industrial expansion..... slaves had indeed been employed (in arts and crafts) before, but the sudden growth of trade....their employment in numbers that would otherwise have been unnecessary. Romans engaged more freely and more openly in various forms of commercial and industrial venture. Yet, even so the agent became more important, for commercial activities became more widespread; and such agents were almost necessarily slaves..... (this is so) because the bonds of slavery (are elastic). They could be so relaxed as to offer an incentive (to the slave) to work by the prospect of wealth and freedom, and so tightened as to provide a guarantee to the master against loss from the misconduct of his slave. In business contracts between slave and master third person seem to have been common, and the work thus done, and no doubt, the profits were considerable........ Renting of land to the slave has already been noticed.... and in industry much the same system was used in various forms; the master might lease a bank, or a business of the use of a ship, the terms being a fixed return or the slave being paid on a commission basis".

"The earnings of the slave became in law his peculium. Once the peculium was saved it might be used to a variety of purposes. No doubt in many cases this fund was expended in providing food or pleasure...... But peculium must not be regarded merely as petty savings, casually earned and idly spent. The slave who made his master's business yield profits, to his own profit too, very often, had a keen sense of the best use to make up his own money. Often he reinvested it in his master's business or in enterprises entirely unrelated to it. He could enter into business relations with hi master, from whom he came to be regarded as entirely distinct, or he could make contracts with a third person. He could even have procurators to manage his own property and interests. And so with the peculium may be found not only land, houses, shops but rights and claims.

"The activities of slaves in commerce are innumerable; numbers of them are shopkeepers selling every variety of food, bread, meat, salt, fish, wine vegetables, beans, Aupine-seed, honey, curd, ham, ducks and fresh fish, others deal inclothing—sandals, shoes, gowns and mantles. In Rome, they plied their trade in the neighbourhood of the Circus Maximus, or the Portions Trigeminus; or the Esquiline Market, or the Great Mart (on the Caolian Hill) or the Suburra.

The extent to which slave secretaries and agents acted for their masters is shown very clearly in the receipts found in the house of Caecilius Jucundus at Pompei.

That the State should possess slaves is not surprising; war, after all, was the affair of the State and the captive might well be State-property. What is surprising is the remarkable use made of public slaves under the Empire and the extraordinary social position occupied by them.....

"Public slave came to mean before the Empire a slave of the state employed in its many offices, and the term implied a given occupation and often social position. The work of slaves of the State, slaves of the townships, and slaves of Caesar comprises much of what would now fall to parts of the higher and the whole of the lower branches of the civil services and of the servants of Municipal Corporations, working both with head and hands... In the subordinate levels (of the Treasury) there worked numbers of clerks and financial officers, all freedmen and slaves. The business dealt with must have been of vast range.... The Mint... the immediate head was a knight, in charge of the minting processes.... a freedman was placed under him, served freedmen and slaves.... From one branch of State service, at any rate, slaves were rigorously excluded, except on one or two occasions of exceptional stress. They were not allowed to fight in the Army because they were not thought worthy of honour. Doubtless other motives were present also; it would be dangerous experiment to train too many slaves systematically in the use of Arms. If, however, slaves served merely in the fighting line, they are regularly to be found in great numbers behind it employed as servants, and in the commissariat and transport. In the fleet slaves were common enough"

Such was the defacto position of the slave in Roman Society. Let us trun to the *defacto* position of the Negro in the United States during the period in which he was slave in the eye of the law. Here are some facts which shed a good deal of light on his position:

Lafayette himself had observed that white and black seamen and soldiers had fought and messed together in the Revolution without bitter difference. Down in Granville Country, North Carolina, a full blooded Negro, John Chavis, educated in Princeton University, was conducting a private school for white students and was a licentiate under the local Presbytary, preaching to white congregations in the State. One of his pupils became Governor of North Carolina, another the State's most prominent Whig senator. Two of his pupils were sons of the Chief Justice of North Carolina. The father of the founder of the greatest military academy of the State attended his school and boarded in his home....

Slave labour was used for all kinds of work and the more intelligent of the Negro slaves were trained as artisans to be used and leased. Slave artisans would bring twice as much as an ordinary field hand in the market. Master craftsmen owned their staff. Some masters, as the system became more involved, hired slaves to their slave artisans. Many slave artisans purchased their freedom by the savings allowed them above the normal labour expected."

The advertisements for runaways and sales are an index to this skill. They received the same or better wages than the poor white labourer and with the influence of the master got the best jobs. The Contractors for masons' and carpenters' work in Athens, Georgia in 1838 were petitioned to stop showing preference to Negro labourers. "The white man is the only real, legal, moral, and civil proprietor of this country and state. The right of his proprietorship reached from the date of the studies of those whitemen. Copernicus and Galileo, who indicated the sphericity of the earth; which sphericity hinted to another white man, Columbus, the possibility by a westerly course of sailing, of finding land.

Hence by whitemen alone was this continent discovered, the whitemen alone, aye, those to whom you decline to give money for bread or clothes for their famishing families, in the logical manner of withholding work from them defending Negroes too in the bargain." In Atlanta in 1858, a petition signed by 2 white mechanics and labourers sought protection against the black slave artisans of masters who resided in other sections. The very next year sundry white citizens were aggrieved that the City Council tolerated a Negro dentist to remain and operate in their midst. 'Injustice to ourselves and the community it ought to be abated. We, the residents of Atlanta, appeal to you for justice'. A Census of free Negroes in Richmond County, Georgia, in 1819 showed carpenters, barbers, boatcorkers, saddlers, spinners, millwrights, holsters, weavers, harness makers, sawmill attendants and steamboat pilots. A Negro shoemaker made by hand the boots in which President Munrow was inaugurated. Harriet Martineau marvelled at the slave workmanship in the delicately tiled floors of Thomas Jefferson's home at Monticello. There still stands in the big house of the old plantation, heavy marks of the hands of these Negro craftsmen, strong mansions built of timber hewn from the original oak and pinned together by wooden pins. Negro women skilled in spinning and weaving worked in the mills. Buckingham in 1839 found them in Athens. Georgia, working alongside with white girls without apparent repugnance of objection.

Negro craftsmen in the South, slave and free fared better than their brothers in the North. In 1856 in Philadelphia, of

1637 Negro craftsmen recorded, less than two-thirds could use their trades; ' because of hostile prejudice '. The Irish who were pouring into America from the very beginning of the nineteenth century were being used in the North on approximately the same motives of preference which governed Negro slavery. 'An Irish Catholic', it was argued in their favour, ' seldom attempts to rise to a higher condition than that in which he is placed, while the Negro often makes the attempt with success.

Had not the old Puritan Oliver Cromwell, while the traffic in black slaves was on, sold all the Irish not killed in the Drogheda Massacre into Barbados? ' Free and fugitive Negroes in New York and Pennsylvania were in constant conflict with this group and the bitter hostility showed itself most violently in the draft riots of the New York. These Hibernians controlled the load carrying and the common labour jobs, opposing every approach of the Negro as a menace to their slight hold upon America and upon a means of livelihood."

Such was the *de facto* condition of the Roman slave and the American Negro slave. Is there anything in the condition of the Untouchables of India which is comparable with the condition of the Roman slave and the American Negro slave? It would not be unfair to take the same period of time for comparing the condition of the Untouchables with that of the slaves under the Roman Empire. But I am prepared to allow the comparison of the condition of the slaves in the Roman Empire to be made with the condition of the Untouchables of the present day.

It is a comparison between the worst of one side and the best of the other, for the present times are supposed to be the golden age for the Untouchables. How does the *defacto* condition of the Untouchables compare with the *defacto* condition of the slaves? How many Untouchables are engaged as the slaves in Rome were, in professions such as those of Librarians, Amanuenses, Shorthand writers? How many Untouchables are engaged, as the slaves in Rome were, in such intellectual occupations as those of rhetoricians, grammarians, philosophers, tutors, doctors and artists? How many untouchables are engaged in trade, commerce or industry as were the slaves in Rome? Even comparing his position with that of the Negro while he was a slave it cannot be said that the condition of the Untouchable has been better. Is their any instance of untouchables having been artisans? Is there any instance of untouchable having maintained a school where Brahmin children have come to sit at his feet in search of learning? Why such a thing is unthinkable? But it has happened in the United States of America. In comparing the *defacto* condition of the Roman slave and the American Negro I have purposely taken the recent condition of the Untouchables as a basis of comparison for the simple reason that the present times are supposed to be the golden age for the untouchables. But comparing even the condition of the untouchables in modern times they are certainly a sunken community as compared with the condition of slaves in time which historians call barbarous.

There can therefore, be no doubt that untouchables have been worse off than slaves. This of course means that untouchability is more harmful to the growth of man than slavery ever was. On this there is a paradox. Slaves who were worse off in law than the untouchables were in fact better off than untouchables and untouchables who were better off in law than slaves were worse off in fact than slaves. What is the explanation of this paradox? The question of all questions is this; what is it which helped the slave to overcome the rigorous denial of freedom by law and enabled them to prosper and grow? What is it that destroyed the effect of the freedom which the law gave to the untouchables and sapped his life of all vitality and stunted his growth.

The explanation of this paradox is quite simple. It will be easily understood if one bears in mind the relation between law and public opinion. Law and public opinion are two forces which govern the conduct of men. They act and react upon each other. At times law goes ahead of public opinion and checks it and redirects in channels which it thinks proper. At times public opinion is ahead of the law. It rectifies the rigour of the law and moderates it. There are also cases where law and public opinion are opposed to each other and public opinion being the stronger of the two forces, disregards or sets at naught what the law-prescribes. Whether through compulsion arising out of convenience of commerce and industry or out of the selfish desire to make the best and the most profitable use of the slaves or out of considerations of humanity, public opinion and law were not in accord with regard to the position of the slave either in Rome or in the United States. In both places the slave was not a legal person in the eye of the law. But in both places he remained a person in the sense of a human being in the eye of the society. To put it differently the personality which the law withheld from the slave was bestowed upon him by society. There lies a profound difference between slavery and untouchability. In the case of the untouchable just the opposite has happened. The personality which the law bestowed upon the untouchables is withheld by society.

In the case of the slave the law by refusing to recognise him as a person could do him no harm because society recognised him more amply than it was called upon to do. In the case of the untouchables the law by recognising him as a person failed to do him any good because Hindu society is determined to set that recognition at naught. A slave had a personality which counted notwithstanding the command of the law. An untouchable has no personality in spite of the command of the law. This distinction is fundamental. It alone can explain the paradox— the social elevation of the slave loaded though he was with the burden of legal bondage and the social degradation of the untouchable aided as he has been with the advantages of legal freedom.

Those who have condemned slavery have no doubt forgotten to take into consideration the fact that in a sense slavery was an apprenticeship in a business, craft or art, albeit compulsory. Unmitigated slavery with nothing to compensate the loss of freedom is of course to be condemned. But to enslave a person and to train him is certainly better than a state of barbarity accompanied by freedom. Slavery did mean an exchange of semi-barbarism for civilisation, a vague enough gift but none the less real. The full opportunities for civilised life could only be fully used in freedom, no doubt, but slavery was an apprenticeship, or in the words of Prof. Myres "an initiation into a higher culture".

This view of slavery is eminently a correct view. This training, this initiation of culture was undoubtedly a great benefit to the slave. Equally it involved considerable cost to the master to train his slave, to initiate him into culture. "There can have been little supply of slaves, educated or trained, before enslavement.

The alternative was to train them when young slaves in domestic work or in skilled craft, as was indeed done to some extent before the Empire, by Cato, the Elder, for example. The training was done by his owner and his existing staff indeed the household of the rich contained special pedagogy for this purpose. Such training took many forms: industry, trade, arts and letter".

The question is why was the slave initiated into the high culture and why did it not fall to the lot of the untouchable to be so initiated? The question is very pertinent and I have raised it because the answer to the question will further reinforce the conclusion that has been reached namely that untouchability is worse than slavery and that is because the slave had a personality and the untouchable has not.

The reason why the master took so much trouble to train the slave and to initiate him in the higher forms of labour and culture was undoubtedly the motive of gain. A skilled slave as an item was more valuable than an unskilled slave. If sold he would fetch better price, if hired out he would bring in more wages. It was therefore an Investment to the owner to train his slave. But this is not enough to account for the elevation of the slave and the degradation of the untouchable. Suppose Roman society had an objection to buy vegetables, milk, butter, water or wine from the hands of the slave?

Suppose Roman society had an objection to allow slaves to touch them, to enter their houses, travel with them in cars, etc. would it have been possible for the master to train his slave, to raise him from semi-barbarism to a cultured state? Obviously not. It is because the slave was not held to be an untouchable that the master could train him and raise him. We again come back therefore, to the same conclusion-namely, that what has saved the slave is that his personality was recognised by society and what has ruined the untouchable is that Hindu society did not recognise his personality, treated him as unfit for human association and common dealing.

That the slave in Rome was no less of a man because he was a slave, that he was fit for human intercourse although he was in bondage is proved by the attitude that the Roman Religion had towards the slave. As has been observed:

"Roman religion was never hostile to the slave. It did not close the temple doors against him; it did not banish him from its festivals. If slaves were excluded from certain ceremonies, the same may be said of free men and women-being excluded from the rites of Bono Dea, Vesta and Ceres, women Jrom those of Hercules at the Ara Maxima. In the days when the old Roman divinities counted for something, the slave came to be informally included in the family, and could consider himself under the protection of the gods of the household.......Augustus ordered that freed women should be eligible as priestesses of Vesta. The law insisted that a slave's grave should be regarded as sacred and for his soul Roman mythology provided no special heaven and no particular hell. Even Juvenal agrees that the slave, soul and body is made of the same stuff as his master."

Slave in Law

There was no stigma attached to his person. There was no gulf social or religious which separated the slave at any rate in Rome from the rest of the society. In outward appearance he did not differ from the free man; neither colour nor clothing revealed his conditions; he witnessed the same games as the freemen, he shared in the life of the Municipal towns, and employed in state service, engaged himself in trade and commerce as all free men did. Often apparent equality in outward things counts far more to the individual than actual identity of rights before the law. Between the slave and the free, there seems often to have been little social barrier. Marriage between slave and freed slave was very common. The slave status carried no stigma on the man in the society. He was touchable and even respectable.

Enough has been said to show that untouchability is worse than slavery. The only thing that is comparable to it is the case of the Jews in the middle ages. The servility of the Jews does resemble to some extent the condition of the untouchables. But there is this to be said about it. Firstly the discrimination made against the Jews was made upon a basis which is perfectly understandable though not justifiable. It was based upon the Jews obstinacy in the matter of religion. He refused to accept the religion of the gentiles and it is his obstinacy which brought about those penalties. The moment he gave up his obstinacy he was free from his disabilities.

This is not the case with the untouchable. His disabilities are not due to the fact that he is a protestant or nonconformist. The second thing to be said about these disabilities of the Jews is that the Jews preferred them to being completely assimilated and lost in the Gentiles. This may appear strange but there are facts to prove it. In this connection reference may be made to two instances recorded in history which typify the attitude of the Jews. The first instance relates to the Napoleonic regime. After the National Assembly of France had agreed to the declaration of the Rights of Man to the Jews, the Jewish question was again reopened by the guild merchants and religious reactionaries of Alsace. Napoleon resolved to submit the question to the consideration of the Jews themselves.

"He convened an Assembly of Jewish Notables of France, Germany and Italy in order to ascertain whether the principles of Judaism were compatible with the requirements of citizenship as he wished to fuse the Jewish element with the dominant population. The Assembly, consisting of I II deputies, met in the Town Hall of Paris on 25th July, 1806, and was required to frame replies to twelve questions relating mainly to the possibility of Jewish patriotism, the permissibility of intermarriage between Jew and non-Jew, and the legality of usury. So pleased was Napoleon with the pronouncements of the Assembly that he summoned a Sanhedrin after the model of the ancient council of Jerusalem to convert them into the decrees of a legislative body.

The Sanhedrin, comprising 71 deputies from France, Germany, Holland and Italy, met under the presidency of Rabbi Sinzheim of Strassburg on 9th February 1807, and adopted a sort of charter which exhorted the Jews to look upon France as their father land, to regard its citizens as their brethren, and to speak its language, and which also pressed toleration of marriages between Jews and Christians while declaring that they could not be sanctioned by the synagogue". It will be noted the Jews refused to sanction intermarriages between Jews and non-Jews. They only agreed to tolerate them. The second instance related to what happened when the Batavian Republic was established in 1795. The more energetic members of the Jewish community pressed for the removal of many disabilities under which they laboured. "But the demand for the full rights of citizenship made by the progressive Jews was at first, strangely enough, opposed by the leaders of the Amsterdam community, who feared that civil equality would militate against the conservation of Judaism and declared that their coreligionists renounced their rights of citizenship in obedience to the dictates of their faith.

This shows that the Jews preferred to live as strangers rather than as members of the community. It is as an 'eternal people' that they were singled out and punished. But that is not the case with the untouchables. They too are in a different sense an "eternal people" who are separate from the rest. But this separateness is not the result of their wish. They are punished not because they do not want to mix. They are punished because they want to.

Untouchability is worse than slavery because slave has personality in the Society while the untouchable has no personality has been made abundantly clear. But this is not the only ground why untouchability is worse than slavery. There are others which are not obvious but which are real nonetheless.

Of these the least obvious may be mentioned as the first. Slavery, if it took away the freedom of the slave, it imposed upon the master the duty to maintain the slave in life and body. The slave was relieved of all responsibility in respect of his food, his clothes and his shelter. All this the master was bound to provide. This was of course no burden because the slave earned more than his keep. But a security for board and lodging is not always possible for every freeman as all wage earners now know to their cost. Work is not always available even to those who are ready to toil but a workman cannot escape the rule according to which he gets no bread if he finds no work. This rule, no work no bread, the ebbs and tides of business, the booms and depression are vicissitudes through which all free wage earners have to go.

But they do not affect the slave who is free from them. He gets his bread-perhaps the same bread, but bread-whether it is boom or whether it is depression. Untouchability is worse than slavery because it carries no such security as to livelihood as the latter does. No one is responsible for the feeding, housing and clothing of the untouchable. From this point of view untouchability is not only worse than slavery but is positively cruel as compared to slavery.

In slavery the master has the obligation to find work for the slave. In a system of free labour workers have to compete with workers for obtaining work. In this scramble for work what chances has the untouchable for a fair deal? To put it shortly in this competition with the scales always weighing against him by reason of his social stigma he is the last to be employed and the first to be fired. Untouchability is cruelty as compared to slavery because it throws upon the untouchables the responsibility for maintaining without any way of earning his living, From another aspect also untouchability is worse than slavery. The slave was property and that gave the slave an advantage over a free man. Being valuable, the master out of sheer self "interest, took great care of the health and well being of the slave.

In Rome the slaves were never employed on marshy and malarial land. On such a land only freemen were employed. Cato advises Roman farmers never to employ slaves on marshy and malarial land.

This seems stranger. But a little examination will show that this was quite natural. Slave was valuable property and as such a prudent man who knows his interest must not expose him to the ravages of malaria. The same care need not be taken in the case of free man because he is not valuable property. This consideration resulted to the great benefit of the slave. He was cared for as no one was. This consideration is completely absent in the case of the untouchable. He is neglected and left to starve and die.

The second or rather the third difference between untouchability and slavery is that slavery was never obligatory. But untouchability is obliged. A person is "permitted" to hold another as his slave. There is no compulsion on him if he does not want to. A Hindu on the other hand is "enjoined" to hold another as untouchable. There is compulsion on the Hindu which he cannot escape whatever his personal wishes in the matter may be.

An Object Surrender

Congress Beats An Inglorious Retreat

The Poona Paet was signed on the 24th September 1932. On 25th September 1932, a public meeting of the Hindus was held in Bombay to accord to it their support. At that meeting the following resolution was passed:

This Conference confirms the Poona agreement arrived at between the leaders of the Caste Hindus and Depressed Classes on September 24, 1932, and trusts that the British Government will withdraw its decision creating separate electorates within the Hindu community and accept the agreement in full.

The Conference urges that immediate action be taken by Government so as to enable Mahatma Gandhi to break his fast within the terms of his vow and before it is too late. The Conference appeals to the leaders of the communities concerned to realize the implications of the agreement and of this resolution and to make earnest endeavour to fulfil them.

This Conference resolves that henceforth, amongst Hindus, no one shall be regarded as an Untouchable by reason of his birth, and that those who have been so regarded hitherto will have the same right as other Hindus in regard to the use of public wells, public schools, public roads, and all other public institutions. This right shall have statutory recognition at the first opportunity and shall be one of the earliest Acts of the Swaraj Parliament, if it shall not have received such recognition before that time.

It is further agreed that it shall be the duty of all Hindu leaders to secure, by every legitimate and peaceful means, an early removal of all social disabilities now imposed by custom upon the so-called Untouchable Classes, including the bar in respect of admission to temples.

This resolution was followed by a feverish activity on the part of the Hindus to throw open. Temples to the Untouchables. No week passed in which the Harijan a weekly paper started by Mr. Gandhi which did not publish a long list of temples thrown open, wells thrown open and schools thrown open to the Untouchables set out under special column headed "Week to Week" on the first page. As samples I produce below these "Week to Week" columns from two issues from the Harijan.

'Harijan' of 18th February 1933

Sri. V. R. Shinde, president, All – India Anti-Untouchability League and Founder – Trustee of the Depressed Mission Society of India, Poona, has addressed an open letter to the members of the Legislative Assembly on Sri. Ranga Iyer's Untouchability Bills, strongly urging them to support the two measures.

In Taikalwadi in 'G' Ward of Bombay, there was an outbreak of fire recently which caused very serious damage to the huts and belongings of 48 Mahar families, The President of the Bombay Provincial Board of the Servants of Untouchables Society sanctioned Rs.500/-for giving relief to these families, and the relief was organized by a sub committee of the 'G" Ward Committee of the Society. A sum of Rs. 402-8 was distributed as an urgent measure of help to the 48 families, containing in all 163 persons.

The Bombay Government has issued orders that requests from local bodies for assignment of Government lands for wells, tanks, Dharamshalas, etc., should not be granted except on condition that all castes alike will have equal use of such wells, tanks, etc.

"Harijan" of July 15, 1933

Week to Week

Educational Facilities

Three reading rooms for Harijans have been opened in the North Arcot District by the S.U.S. In the Madura District S.U.S. workers got Harijan children admitted into the Viraganur taluq board school. Banians, towels, slates, etc. were distributed free to the children of the Melacheri school established by the Madura S.U.S Two Harijan, students of Ramjas College, Delhi, have been allowed free scholarship and free lodging and one a free scholarship by Principal Thadani of the College. One night school for adult Harijans was opened under the auspices of the Lahore Harijan Seva Sangh in the Harijan quarters outside. Mochi Gate. The opening ceremony was performed by Mrs. Brij Lal Nehru. It has been decided to start one more hostel for Harijan students in Brahmana Kodur (Guntur). The

East Godavery District Harijan Seva Sangham has resolved to start a hostel for Harijan Girl Students studying in Coconada. A sum of Rs. 630/-20 bags of rice, fuel necessary for one year, have been already received as donations for the hostel, which will be started with 15 students. The Anantapur District Harijan Seva Sangam has decided to start a hostel for Harijan students in Uravakonda. Some provisions and money have already been collected and it is intended to start the hostel with 20 students. Owing to the unremitting efforts of the District Harijan Seva Sangham, Guntur, Harijan boys have been allowed into the savarna schools in a manner of villages and towns.

Wells

Three wells in Coimbatore District which were in a bad condition, were cleaned and made available for use. The District Board President, South Arcot, has promised to dig four wells in cherries selected by the S.U.S During the fortnight ending 31-5-33, no less than 125 wells in all were opened to Harijans and 5 new ones constructed in Andhradesh.

General

A shop has been opened in a bustee near Hogg Market (Calcutta) where Doms live, for supplying them with articles of food at cheap rates. Rs. 60 has been paid by the S.U.S. Bengal for paying up the debts of a Harijan family at Bibi Bagan bustee (Calcutta). The Amrita Samaj (Calcutta) has given service to some Harijans. 450 Harijans of Bolpur (Birbhum) have given up drinking habits and 1,275 Muchis have taken a vow not to take beef. Three new district centres of S.U.S have been opened during the month in Bankura, Murshidabad, and 24 Parganas. Trichinopoly, Tanjore, Tinnevelley, Salem, Dindigal, North Arcot and Madura have all taken up the idea of a Gandhi Harijan Service corps for direct and personal service in the cheris.

Alandural, a Harijan village 12 miles from Coimbatore was given Rs. 25/-worth of grain, Rs. 100 worth of cloth and Rs. 5 worth of oil, as relief after a fire in the village. A Harijan Youth League has been formed in Chindambaram. A shop to supply provisions at cost price to the Harijans has been set up in Tenali and is being made use of by them. A sum of Rs.110/-

was spent in giving help for rebuilding houses of Harijans in Valanna Palem (East Kistna) recently destroyed by fire. A sum of Rs.100/-was contributed by the Provincial Committee towards the relief of Harijans in Yellamanchili (Vizag) who lost their houses by a fire. The local Harijan Seva Sangham is endeavouring to erect new houses for the Harijans in a better locality and is collecting donations in cash and building materials. One Harijan has been employed as a servant by a Savarna Gentleman in Gollapalem.

When the owners or trustes of temples were not prepared to throw open their temples to the Untouchables, the Hindus actually started Satyagrah against them to compel them to fall in line. The Satyagrah by Mr. Kelappan for securing entry to the Untouchables in the temple at Guruvayur was a part of this agitation. To force the hands of the trustees of the temples who had the courage to stand against the current, many Hindu legislators came forward, tumbling over one another, with Bills requiring the trustees to throw open temples to the Untouchables if a referendum showed that the majority of the Hindu worshippers voted in favour. There was a spate of such Bills and a race among legislators to take the first place. There was a Temple Entry Bill by Dr. Subbaroyan of the Madras Legislative Council. There were four Bills introduced in the Central Assembly. One was by Mr. C.S. Ranga Iyer, another by Mr. Harabilas Sarda, a third by Mr. Lalchand Navalrai, and a fourth one by Mr. M.R. Jayakar.

In this agitation Mr. Gandhi also joined. Before 1932, Mr. Gandhi was opposed to allow Untouchables to enter Hindu Temples. To quote his own words Mr. Gandhi said:

> *'How is it possible that the Antyajas (Untouchables) should have the right to enter all the existing temples? As long as the law of caste and Ashram has the chief place in Hindu Religion, to say that every Hindu can enter every temple is a thing that is not possible today."*

His joining the movement for Temple entry must therefore remain a matter of great surprise. Why Mr. Gandhi took this some sault it is difficult to imagine. Was it an honest act of change of heart, due to a conviction that he was in error in opposing the entry of the Untouchables in Hindu temples? Was it due to a realization that the political separation between the Hindus and the Untouchables brought about by the Poona Pact might lead to a complete severance of the cultural and religious ties and that it was necessary to counteract the tendency by some such measure as Temple Entry as will bind the two together? Or was his object in joining the Temple Entry movement to destroy the basis of the claim of the Untouchables for political rights by destroying the barrier between them and the Hindus which makes them separate from the Hindus? Or was it because Mr. Gandhi saw before him looming large a possibility of adding to his name and fame and rushed to make the most of it, as is his habit to do? The second or the third explanation may be nearer the truth.

What was the attitude of the Untouchables to this movement for Temple entry? I was asked by Mr. Gandhi to lend my support to the movement for Temple entry. I declined to do so and issued a statement on the subject to the Press. As it will help the reader to know the grounds for my attitude to this question I have thought it well to set it in full. Here it is!

Statement on Temple Entry Bill

14th February, 1933

Although the controversy regarding the question of Temple Entry is confined to the Sanatanists and Mahatma Gandhi, the Depressed Classes have undoubtedly a very important part to play in it, in so far as their position is bound to weigh the scales one way or the other when the issue comes up for a final settlement. It is, therefore, necessary that their view point should be defined and stated so as to leave no ambiguity about it,

To the Temple-Entry of Mr. Ranga Iyer as now drafted, the Depressed Classes cannot possibly give their support. The principle of the Bill is that if a majority of Municipal and Local Board voters in the vicinity of any particular temple on a referendum decide by a majority that the Depressed Classes shall be allowed to enter the temple, the Trustees or the Manager of that temple shall give effect to that decision. The principle is an ordinary principle of Majority rule, and there is nothing radical or revolutionary about the Bill, and if the Sanatanists were a wise lot, they would accept it without demur.

The reasons why the Depressed Classes cannot support a Bill based upon this principle are two: One reason is that the Bill cannot hasten the day of temple-entry for the Depressed Classes any nearer than would otherwise be the case. It is true that under the Bill, the minority will not have the right to obtain an injunction against the Trustee, or the Manager who throws open the temple to the Depressed Classes in accordance with the decision of the majority. But before one can draw any satisfaction from this clause and congratulate the author of the Bill, one must first of all feel assured that when the question is put to the vote there will be a majority in favour the Temple Entry. If one is not suffering from illusions of any kind one must accept that the hope of a majority voting in favour of Temple-entry will be very rarely realized, if at all. Without doubt, the majority is definitely opposed today – a fact which is conceded by the author of the Bill himself in his correspondence with the Shankaracharya.

What is there in the situation as created after the passing of the Bill, which can lead one to hope that the majority will act differently? I find nothing. I shall, no doubt, be reminded of the results of the referendum with regard to the Guruvayur Temple. But I refuse to accept a referendum so over weighted as it was by the life of Mahatma Gandhi as the normal result. In any such calculations, the life of the Mahatma must necessarily be deducted.

Secondly, the Bill does not regard Untouchability in temples as a sinful custom. It regards Untouchability merely as a social evil not necessarily worse than social evils of other sorts. For, it does not declare Untouchability as such to be illegal. Its binding force is taken away, only, if a majority decides to do so. Sin and immorality cannot become tolerable because a majority is addicted to them or because the majority chooses to practise them. If Untouchability is a sinful and an immoral custom, then in the view of the Depressed Classes it must be destroyed without any hesitation even if it was acceptable to the majority. This is the way in which all customs are dealt with by Courts of Law, if they find them to be immoral and against public policy.

This is exactly what the Bill does not do. The author of the Bill takes no more serious view of the custom of Untouchability than does the temperance reformer of the habit of drinking. Indeed, so much is he impressed by the assumed similarity between the two that the method he has adopted is a method which is advocated by temperance reformers to eradicate the evil habit of drinking, namely, by local option. One cannot feel much grateful to a friend of the Depressed Classes, who holds Untouchability to be no worse than drinking. If Mr. Ranga Iyer had not forgotten that only a few months ago Mahatma Gandhi had prepared himself to fast unto death if Untouchability was not removed, he would have taken a more serious view of this curse and proposed a most thorough going reform to ensue its removal lock, stock and barrel. Whatever its shortcomings may be front the stand point of efficacy, the least that the Depressed classes could expect is for the Bill to recognize the principle that Untouchability is a sin.

I really cannot understand how the Bill satisfies Mahatma Gandhi who has been insisting that Untouchability is a sin. It certainly does not satisfy the Depressed Classes. The question whether this particular Bill is good or bad, sufficient or insufficient, is a subsidiary question.

The main question is; do the Depressed Classes desire Temple Entry or do they not? This main question is being viewed by the depressed Classes by two points of view. One is the materialistic point of view. Starting from it, the Depressed Classes think that the surest way for their elevation lies in higher education, higher employment and better ways of earning a living. Once they become well placed in the scale of social life, they would become respectable and once they become respectable the religious outlook of the orthodox towards them is sure to undergo change, and even if this did not happen, it can do no injury to their material interest. Proceeding on these lines the Depressed Classes say that they will not spend their resources on such an empty thing as Temple Entry.

There is also another reason why they do not care to fight for it. That argument is the argument of self-respect.

Not very long ago there used to be boards on club doors and other social resorts maintained by Europeans in Indian, which said "Dogs and Indians" not allowed. The temples of Hindus carry similar boards today, the only difference is that the boards on the Hindu temples practically say: "All Hindus and all animals including dogs are admitted, only Untouchables not admitted." The situation in both cases is on a parity. But Hindus never begged for admission in those places from which the Europeans in their arrogance had excluded them. Why should an Untouchable beg for admission in a place from which he has been excluded by the arrogance of the Hindus? This is the reason of the Depressed Class man who is interested in his material welfare. He is prepared to say to the Hindus, "to open or not to open your temples is a question for you to consider and not for me to agitate. If you think, it is bad manners not to respect the sacredness of human personality, open your temples and be a gentleman. If you rather be a Hindu than be gentleman, then shut the doors and damn yourself for I do not care to come."

I found it necessary to put the argument in this form, because I want to disabuse the minds of men like Pandit Madan Mohan Malaviya of their belief that the Depressed Classes are looking forward expectantly for their patronage.

The second point of view is the spiritual one. As religiously minded people, do the Depressed Classes desire temple entry or do they not? That is the question. From the spiritual point of view, they are not indifferent to temple entry as they would be, if the material point of view alone were to prevail. But their final answer must depend upon the reply which Mahatma Gandhi and the Hindus give to the question namely: What is the drive behind this offer of temple entry? Is temple entry to be the final goal of the advancement in the social status of the Depressed Classes in the Hindu fold? Or is it only the first step and if it is the first step, what is the ultimate goal? Temple Entry as a final goal, the Depressed Classes can never support. Indeed they will not only reject it, but they would then regard themselves as rejected by Hindu society and free to find their own destiny elsewhere. On the other hand, if it is only to be a first step in the direction they be may be inclined to support it.

The position would then be analogous to what is happening in the politics of India today. All Indians have claimed Dominion Status for India. The actual constitution will fall short of Dominion Status and many Indians will accept it. Why? The answer is that as the goal is defined, it does not matter much if it is to be reached by steps and not in one jump. But if the British had not accepted the goal of Monition Status, no one would have accepted the partial reforms which many are now prepared to accept. In the same way, if Mahatma Gandhi and the reformers were to proclaim what the goal which they have set before themselves is for the advancement of the social Status of the Depressed Classes in the Hindu fold, it would be easier for the Depressed Classes to define their attitude towards Temple Entry. The goal of the depressed Classes might as well be stated here for the information and consideration of all concerned. What the Depressed Classes want is a religion, which will give them equality of social status.

To prevent any misunderstanding, I would like to elaborate the point by drawing a distinction between social evils which are the results of secular causes and social evils which are founded upon the doctrine of religion. Social evils can have no justification whatsoever in a civilized society. But nothing can be more odious and vile than that admitted social evils should be sought to be justified on the ground of religion. The Depressed Classes may not be able to overthrow inequities to which they are being subjected. But they have made up their mind not to tolerate a religion that will lend its support to the continuance of these inequities.

If the Hindu religion is to be their religion, then it must become a religion of Social Equality. The mere amendment of Hindu religious code by the mere inclusion in it of a provision to permit temple entry for all, cannot make it a religion of equality of social status. All that it can do is to recognize them as nationals and not aliens, if I may use in this connection terms which have become so familiar in politics. But that cannot mean that they would thereby reach a position where they would be free and equal, without being above or below any one else, for the simple reason that the Hindu religion does not recognize the principle of equality of social status; on the other hand it fosters inequality by insisting upon grading people as Brahmins, Kshatrias, Vaishyas and Sudras, which now stand towards one another in an ascending scale of hatred and descending scale of contempt. If the Hindu religion is to be a religion of social equality then an amendment of its code to provide temple-entry is not enough. What is required is to purge it of the doctrine of Chaturvarna. That is the root cause of all in equality and also the parent of the caste system and Untouchability, which are merely forms of inequality. Unless it is done not only will the Depressed Classes reject Temple Entry, they will also reject the Hindu faith. Chaturvarna and the Caste system are incompatible with the self-respect of the Depressed Classes. So long as they stand to be its cardinal doctrine the Depressed Classes must continue to be looked upon as low. The Depressed Classes can say that they are Hindus only when the theory of Chaturvarna and caste system is abandoned and expunged from the Hindu shastras.

Do the Mahatma and the Hindu reformers accept this as their goal and will they show the courage to work for it? I shall look forward to their pronouncements on this issue, before I decide upon my final attitude. But whether Mahatma Gandhi and the Hindus are prepared for this or not, let it be known once for all that nothing short of this will satisfy the Depressed Classes and make them accept Temple Entry. To accept temple entry and be content with it, is to temporize with evil and barter away the sacredness of human personality that dwells in them.

There is, however, one argument which Mahatma Gandhi and the reforming Hindus may advance against the position I have taken. They may say: acceptance by the Depressed Classes of Temple Entry now, will not prevent them from agitating hereafter for the abolition of Chaturvarna and Caste. If that is their view, I like to meet the argument right at this stage so as to clinch the issue and clear the road for future developments.

My reply is that it is true that my right to agitate for the abolition of Chaturvarna and Caste System will not be lost, if I accept Temple Entry now. But the question is on what side will Mahatma Gandhi be at the time when the question is put. If he will be in the camp of my opponents, I must tell him that I cannot be in his camp now. If he will be in my camp he ought to be in it now.—B.R. Ambedkar.

Dewan Bahadur R. Srinivasan who along with me represented the untouchables at the Round Table Conference also did not support the movement for Temple entry. In a statement to the Press, he said:

> *"When a Depressed Classes member is permitted to enter into the caste Hindu temples he would not be taken into any one of the four castes, but treated as man of fifth or the last or the lower caste, a stigma worse than the one to be called an Untouchable. At the same time he would be subjected to so many caste restrictions and humiliations. The Depressed Classes shun the one who enters like that and exclude him as casteman. The crores of Depressed Classes would not submit to caste restrictions. They will be divided into sections if they do. 'Temple entry cannot be forced by law. The village castemen openly or indirectly defy the law. To the village Depressed Class man it would be like a scrap of paper on which the "Sugar" was written and placed in hands for him to taste. The above facts are placed before the public in time to save confusion and disturbance in the country."*

To the question I put to Mr. Gandhi in my statement he gave a straight reply. He said that though he was against untouchability he was not against caste. If at all, he was in favour of it and that he would not therefore carry his social reform beyond removing untouchability. This was enough for me to settle my attitude. I decided to take no further part in it. The only leading member from the Untouchable community was the late Dewan Bahadur Rajah. One cannot help saying that he played a very regrettable part in this business. The Dewan Bahadur was a nominated member of the Central Assembly from 1927. He had nothing to do with the Congress either inside or outside the Assembly. Neither by accident nor by mistake did he appear on the same side of the Congress. Indeed, he was not merely a critic of the Congress but its adversary. He was the staunchest friend of the government and never hesitated to stand by the government. He stood for separate electorates for the Untouchables to which the Congress was bitterly opposed. In the crisis of 1932, the Dewan Bahadur suddenly decided to desert the Government and take sides with the Congress. He became the spearhead of the Congress movement for joint electorates and Temple entry.

It is impossible to discover a parallel in the conduct of any other public cause. The worst part of the business was that it had none but personal motive behind. The Dewan bahadur was deeply cut because the Government did not nominate his as a delegate to the Round Table Conference to represent the Untouchables and in his stead nominated Dewan Bahadur R. Srinivasan. The government of India had good ground for not nominating him. It was decided that neither the members of the Simon Commission not the members of the Central Legislative Committee should have a place in the Round Table Conference. The Dewan Bahadur was a member of the Central Legislative Committee and had therefore to be dropped.

This was quite a natural explanation. But the wounded pride of Dewan Bahadur Rajah could not let him see it. When the Congress Ministry took office in Madras, when he saw how the Poona Pact was being trampled upon, how his rival was made a Minister and how notwithstanding his services to the Congress he was left out, he bitterly regretted what he did. The fact, however, remains that in the critical year of 1932, Dewan Bahadur Rajah lent his full support to the Congress. He was not only running with the Congress crowd but he took care not to fall out in the race for legislation against untouchability. He too had sponsored two Bills. One of them was called the Removal of Untouchability Bill and the other was called the Criminal Procedure Amendment Bill.

Mr. Gandhi did not mind any opposition and was indifferent as to whether it came from the orthodox Hindus or from the Untouchables. He went on in mad pursuit of his object. It is interesting to ask, what happened to this movement? Within the short compass of this book it is not possible to spread out this inquiry and cover everything that was done and claimed as evidence of the success of the movement.

To put it briefly, after a short spurt of activity in the direction of removing untouchability by throwing open temples and wells the Hindu mind returned to its original state. The reports appearing in the "Week to Week" columns of the Harijan subsided, became few and far between and ultimately vanished. For myself I was not surprised to find that the Hindu heart was so soon stricken with palsy. For I never believed that there was so much milk of human kindness locked up in the Hindu breast as the "Week to Week" column in the Harijan would have the world believe. As a matter of fact a large part of the news that appeared in the "Week to Week" was faked and was nothing but a lying propaganda engineered by Congressmen to deceive the world that the Hindus were determined to fight untouchability.

Few temples if any were really opened and those that were reported to have been opened most of them were dilapidated and deserted temples which were used by none but dogs and donkeys. One of the evil effects of the Congress agitation is that it has made the political minded Hindus a lying squad which will not hesitate to tell any lie if it can help the Congress. Thus ended the part which the Hindu public played or was made to appear to play in this Temple-Entry movement. The same fate overtook the Guruvayur Temple Satyagrah and the legislation for securing Temple-Entry for the Untouchables. As these are matters which were pursued by Mr. Gandhi and congressmen their history might be told in some detail inasmuch as it reveals the true mentality of Mr. Gandhi and the Congress towards the Untouchables.

To begin with the Guruvayur Temple Satyagrah. A temple of Krishan is situated at Guruvayur in the Ponnani taluk in Malabar. The Zamorin of Calicut is the trustee of the temple. One Mr. Kelappan, a Hindu who was working for the cause of the Untouchables of Malabar, began an agitation for securing the Untouchables entry into the temple. The Zamorin of Calicut as the trustee of the temple refused to throw open temple to the Untouchables and in support of his action cited Section 40 of the Hindu Religious Endowments Act which said that no trustee could do anything against the custom and usage of the temples entrusted to him.

On the 20th September 1932, Mr. Kelappan commenced a fast in protest lying in front of the temple in the sun till the Zamorin revised his views in favour of the Untouchables. To get rid of this annoyance and embarrassment the Zamorin appealed to Mr. Gandhi to request Mr. Kelappan to suspend his fast for a time. After a fast for ten days Mr. Kelappan at the request of Mr. Gandhi suspended the fast on 1st October 1932 for three months. The Zamorin did nothing. Mr. Gandhi sent him a wire telling him that he must move in the matter and get over all difficulties legal or otherwise. Mr. Gandhi also told the Zamorin that as Mr. Kelappan had suspended his fast on his advice he had become responsible for securing to the Untouchable entry into the temple to the extent of sharing the fast with Mr. Kelappan. On 5th November 1932, Mr. Gandhi issued the following statement to the press:

> *"There is another fast which is a near possibility and that in connection with the opening of the Guruvayur temple in Kerala. It was at my urgent request that Mr..Kelappan suspended his fast for three months, a fast that had well nigh brought him to death's door. I would be in honour bound to fast with him if on or before 1st January 1933 that temple is not opened to the Untouchables precisely on the same terms as to the Touchables, and if it becomes necessary for Mr. Kelappan to resume his fast."*

The Zamorin refused to yield and issue a counter-statement to the press in which he said:

> *"The various appeals that are being made for throwing open the temples to Avarnas proceed upon an inadequate appreciation of such difficulties. In these circumstances, there is hardly any justification for thinking that it is in my power to throw open the Guruvayur temple to the Avarnas as desired by the supporters of the temple-entry campaign"*

In these circumstances a fast by Mr. Gandhi became in evitable, and obligatory. But Mr. Gandhi did not go on fast. He modified his position and said that he would, refrain from fasting if a referendum was taken in Ponnani taluka in which the temple was situated and if the referendum showed that the majority was against the throwing open of the temple to the Untouchables. Accordingly, a referendum was taken. Voting was confined to those who were actual temple goers Those who were not entitled to enter the temple and those who would not enter it were excluded from the voters' list. It was reported that 73 per cent of eligible voters voted. The result of the poll was 56 per cent, were in favour of temple entry, 9 per cent, against, 8 percent were neutral and 27 percent abstained from recording their votes.

On this result of the referendum, Mr. Gandhi was bound to start the fast. But he did not. Instead, on the 29th of December

1932 Mr. Gandhi issued a statement to the press which he concluded by saying:

> *"In view of the official announcement that the Vice regal decision as to sanction for the introduction, in the Madras Legislative Council, of Dr. subbaroyan's permissive Bill with reference to the temple-entry could not possibly be announced before the 15th January, the fast contemplated to take place on the second day of the New Year will be indefinitely postponed and in any case up to the date of the announcement of the Viceregal decision. Mr. Kelappan concurs in this postponement."*

The Viceregal pronouncement mentioned by Mr. Gandhi had reference to the Viceroy's granting permission or refusing permission to the moving of the Temple Entry Bills in the Legislature. That permission was given by the viceroy. Yet Mr. Gandhi did not fast. Not only did he not fast, he completely forgot the matter as though it was of no moment. Since then nothing has been heard about Guruvayur Temple Satyagrah though the Temple remains closed to the Untouchables even today.

Thus ended Guruvayur. Let me now turn to the other project namely legislation for temple-Entry. Of the many bills the one in the name of Mr. Ranga Iyer in the Central Legislature was pursued. The rest were dropped. There was a storm at the very birth of the Bill. Under the Government of India Act as it then stood no legislative measure which affected religion and customs and usages based on religion could be introduced in the Assembly unless it had the previous sanction of the Governor-General. When the Bill was sent for such sanction another commotion was created by the reports that were circulated that the governor-General was going to refuse his sanction. Mr. Gandhi was considerably excited over these reports. In a statement to the press issued on the 21st January 1933, Mr. Gandhi said:

> *"If the report is an intelligent anticipation of the forth coming viceregal decision, I can only say that it will be a tragedy.... I emphatically repudiate the suggestion that there is any political objective behind these measures. If court decisions had not hardened a doubtful custom into law, no legislation would be required. I*

would myself regard State interference in religious matters as an intolerable nuisance. But here legislation becomes an imperative necessity in order to remove the legal obstruction and based as it will be on popular will, as far as I can see, there can be no question of clash between parties representing rival opinions."

The decision of the Government was announced on the 23rd of January 1933. Lord Willingdon refused sanction to Dr. Subbaroyan's Temple-Entry Bill in the Madras Council, but His Excellency permitted the introduction, in the Legislative Assembly, of Mr. Ranga Iyer's Untouchability abolition Bill. The Government emphasized the need of ascertainment of Hindu opinion before they (Government) could decide what attitude to adopt. The announcement further stated that the Governor-General and the Government of India desired to make it plain that it was essential that consideration of any such measure should not proceed unless the proposals were subjected to the fullest examination in all their aspects, not merely in the Legislature but also outside it, by all who would be affected by them. This condition can only be satisfied if the Bill is circulated in the widest manner for the purpose of eliciting public opinion. It must also be understood that the grant of sanction to the introduction in the Central Legislature, Bills relating to temple entry do not commit the Government in any way to the acceptance or support of the principles contained therein. On the next day, Mr. Gandhi issued a statement in which he said:

"I must try to trace the hand of god init. He wants to try me through and through. The sanction given to the all India Bill was an unintentional challenge to Hinduism and the reformer. Hinduism will take care of itself if the reformer will be true to himself. Thus considered the government of India's decision must be regarded as Godsend. It clears the issue. It makes it for India and the world to understand the tremendous importance of the moral struggle now going on in India. But whatever the Sanatanists may decide the movement for Temple-Entry now broadens from Guruvayur in the extreme south to Hardwar in the north and my fast, though it remains further postponed, depends not now

upon ' Guruvayur only but extends automatically to temples in general."

One can well realize under what fanfare the Bill began its legislative career. On the 24th of March 1933, Mr. Ranga Iyer formally introduced the Bill in the assembly. As it was a Bill for Mr. Gandhi the Congress members of the Assembly were of course ready to give it their support. Mr. Gandhi had appointed Mr. Rajagopalachari and Mr. G.D. Birla to canvass support for the Bill among the Non-congress members with a view to ensure safe passage for the Bill. He said they were better lobbyists than he was. The motion for introduction was opposed by the Rajah of Kollengode and Mr. Thampan raised a preliminary objection that the Bill was ultra vires of the legislature.

The latter objection was overruled by the President and the House allowed the Bill to be introduced. Mr. Ranga Iyer next moved that the Temple-Entry Bill be circulated to calicut public opinion by the 30th July. Raja Bahadur Krishnamachari opposed the circulation motion and condemned the proposed legislation in strong terms. At last he urged that the date for circulation should be 31st December instead of 31st July. Mr. Gunjal opposed the circulation motion and asked the House not to support the Bill. As it was already 5 p.m. and as that was the last day of the session for non-official business, the President wanted to take the sense of the House for a late sitting. As there was no over whelming majority for it, the President adjourned the House. So the Bill stood postponed to the autumn session of the Assembly.

The discussion of the Bill was resumed on 24th August 1933 during the Autumn session of the Central Legislature. Sir Harry Haig on behalf of the Government explained that their support to the motion for circulation of the Bill should in no way be construed as implying support to its provisions. It was true that the Government sympathized for the Depressed Classes and were anxious to do what they could for their social and economic improvement. He quoted from the communique issued in January last, wherein the government's view was fully explained. In his opinion circulation by the end of June was a fair and reasonable time to secure the widest possible circulation.

As regards the limit of circulation to temple going Hindus, Sir Harry Haig said from the practical viewpoint that it would really hardly be possible to impose the restriction as proposed. The government wanted the matter to be fully discussed by all classes of Hindus and were therefore prepared to give their support to the amendment of Mr. Sharma. Closure was moved and the House accepted Mr. Sharma's motion for circulation of the Bill by the end of June 1934. Opinions were duly received. They fill a whole volume of over a thousand fullscape pages. The Bill was ready for the next stage namely to move for the appointment of a Select Committee.

Mr. Ranga Iyer had even given notice for such a motion. A strange thing happened. The government of India decided to dissolve the Assembly and order new election. The result of this announcement was a sudden change in the attitude of the Congress members in the Central Legislature towards Mr. Ranga's Bill. One and all stood out against it and refused to give any further support to the Bill. They were terrified of the electorates. Mr. Ranga Iyer's position was very pitiable. He described it in very biting language, the venom of which could hardly be improved upon. So well did he describe the situation that I make no apology for reproducing the following extract from his speech. Rising to move his motion Mr. Ranga Iyer said:

Sir, I rise to move what is known as the Temple-Entry Bill, to remove the disabilities of the so-called Depressed Classes. Sir, I move:

> *"That the Bill to remove the disabilities of the so-called Depressed Classes in regard to entry into Hindu temples be referred to a Select Committee consisting of the Honourable sir Nripendra Sircar, the Honourable Sir Henry Craik, Bhai Parma Naud, Rao Bahadur M.C. Rajah, Mr. T.N. Ramakrshna Reddi, Rao Bahadur B.L. Patil and the Mover,"*

I will delete with your permission, the words 'with instructions to report within a fortnight' and then I will continue the remaining portion of the motion: 'and that the number of members whose presence shall be necessary to constitute a meeting of the Committee shall be five' "Sir, at the time I gave notice of this motion, I did not think that before a fortnight we would be going into the wilderness. Therefore, I recognize the limitations of this motion, for there will be no time even to go to a Select Committee. I recognize that it gives us an opportunity to express our opinion on the subject.

"I have already stated that I owed an apology to Mr. Satyamurthi for while interrupting Mr. Mudaliar, I was not in a position naturally as he was rushing along with his speech to explain myself fully and he would have been at a disadvantage if I had done so.

I recognize that Mr. Satyamurthi, who was at no time in favour of the Temple-entry Bill, has succeeded in making the Congress drop it. I read the following written statement of Mr. C. Rajagopalachari in the Hindu of Madras, dated the 16th august. The Hindu is a very responsible newspaper, and as it is not a mere telegraphic interview but a written statement, I believe Mr. Rajagopalachari's statement can be taken as accurate. Mr. Rajagopalachari is apologizing to the public for his betrayal of the cause of the Untouchables, As the principal lieutenant of Mahatma Gandhi, his betrayal must be placed on record. He says:

> *'The question has been asked by some Sanatanists whether Congress candidates will give an undertaking that Congress will not support any legislative interference with religious observances. Similar questions may be asked on a variety of topics by persons and groups interested in each one of them. That such questions are asked only of the Congress candidates and similar elucidation is not attempted in respect of other parties and independent candidates is a very great compliment paid to the Congress."*

"So, says, Sriman Rajagopalachari, And, instead of following up the compliment and arousing public opinion on an unpopular measure, here is a great Congress leader who sat dharna at our house with his son-in-law, Devidas Gandhi,, who repeatedly called on me at Delhi and said 'We seek joint support for this legislative measure,' here is a man who goes back 'like a crab' tomorrow the language of Shakespeare. Political parties, explains this subtle brain from the south, have distinctive policies on various questions covering a wide field:

"Not all of them, however, are made into election issues at any one time'.

' Sir, this Congress leader is afraid of facing the public opinion which he has roused.

"Sir, are the congress people slaves?

"They are slaves who fear to speak,

For the fallen and the weak.'

"According to Milton, To say and straight unsay argues no liar but a coward traced,' Mr. Rajagopalachari unsays now what he had been saying long before the General Election from every platform in the following words:

"the congress candidates go to the electorate in this election on well-defined political issues."

"That is to say, they go to the electorate with a view to pandering to the prejudice of the masses whom they have misled, so much so, that they have got themselves into a bog. Lord Willingdon came to their rescue, to take them out of the bog by announcing the dissolution of this assembly and giving them an opportunity, as a Constitutional Viceroy, to return to the sheltered paths of constitutionalism. Therefore, they have run away from their own convictions and are playing every trick to come back to the Legislature with as large a number as possible. Had they gone on with the Temple entry Bill or the Untouchability question, they would have lost many votes, for it is not a popular issue. I said so, though Mahatma Gandhi contradicted me publicly at the time, I said so when Shankaracharya was staying in Malabar in my brother's house at Palghat. My brother came on a deputation to the Viceroy to oppose the Bill. I said: ' I know, the reformer is not in a majority in Malabar' Nowhere else are the reformers in a majority but the reformers believe, in persuading the majority to their way of thinking.

Then, I said-whatever the result of a referendum, the Congress people might have taken in guruvayur in Malabar, might be, I could not for a moment believe that the majority of the temple-going people in Malabar were in favour of admitting the Untouchables into the temples: but I was prepared to fight them, also to argue with them and to persuade them and to make them take an interest in the cause and the case of the Untouchables, for, I feel, the Untouchables are a part of my community.

Sir, if one third of my community is to remain submerged in exclusion in the name of religion, I feel, as I have always felt and said, that community has no right to existence. It is with a view to the unification of the Hindu community, it is with a view to building up the greatness of the future of that community on the past of that community, when Untouchability was quite unknown as in the Vedic ages, that I have taken up their cause.

And now, I find Congressmen, so keen about Untouchability yesterday, explaining why they are not taking it up today. Mr. Rajagopalachari has driven the last nail into the coffin of the Temple Entry Bill as Raja Bahadur Krishnamachari, the Raja Saheb of Kollengode or Sir Satya Charan Mukherji would perhaps like to say, representing as they do the various Sanatanist groups of the country.

"Sir. Mr. Rajagopalachari goes on to say that they asked to be returned' on no other issue,' that is to say, not on Temple Entry issue, but merely on a political Anglo-phobia issue, an anti-British issue, because, having traded on public feeling, having tried to give it as much racial antipathy as possible in the name of non-violence, in the name of religion itself, because non-violence was sometimes given a religious bias, having created that atmosphere might not help them in the election if they fought it on a bigger, a cleaner and higher issue, namely, the removal of Untouchability itself, they sidetrack the issue, they run away from their conviction:

'They are slaves who dare not be

In the right with two or three.'

"Then he, a principal lieutenant of Gandhiji goes on to say;

'If successful at the polls, they cannot believe they will receive the mandate of the electorate on any other questions.'

"That is to say, they are not receiving the mandate of the electorate on the Temple Entry Bill.

This man, who came screaming at our doors, begging us for support-these beggars in the cause of the Congress—who just begged of us to proceed with this Temple Entry Bill, are not only betraying the cause of the Untouchables, but they are betraying the principles of the Mahatma himself, for, we know, that Mahatma's fast was directed toward the uplift of the Untouchables by giving them concession in regard to the Communal Award, which the Congress naturally has hesitated to repudiate, and we, therefore, know that has a direct bearing on the Untouchability question to approach which, to solve which, the Mahatma, the great Mahatma, wanted to tour the country, but today the Congress, who betrayed him first in the betrayal of the Congress boycott of the Councils, have, by seeking to come to the Councils, further betrayed him with the assistance of his own Samandhi, Rajagopalachari, and they say that they are not going to proceed with the Untouchability question and the Temple Entry Bill without a mandate from the people.

"Sir, where is the difference, I ask, between Raja Bahadur Krishnamachariar and Sriman Rajagopalachari? Raja Bahadur Krishnamachariar has always conceded – 'take a mandate from the people and then come and legislate,' Sir, he is not a coward; a great Sanatanist himself, he is willing to face the music. On the contrary, these people who pillory the Sanatanists up and down the country, forgetting that Sanatan Dharma is eternal truth itself, are behaving in a manner which even the Sanatanists will not appreciate, for Sanatan Dharma is eternal truth and the betrayal of truth is worthy only of untruthful people. Having betrayed many a principle which would lead us to our national goal, having taken up the case of the Untouchables only to save their faces, with no conviction behind them, as we now see, the great Congress leaders with the exception of Mahatma Gandhi, have said through Rajagopalachari, the Organizer-in-chief of the coming elections on behalf of the Congress;

'It will be open to all Congress men to have the matter duly considered before it is ever made into an official Congress Bill.'

"For this betrayal of the cause of the Untouchables, I hope constitutionalists will organize themselves, whether Hindus or Musalmans. They can agree to differ later on communal issues, but they will unite and offer a great battle to the Congress and bring that organ of masqueraders down on its knees, Sir, I think here is a betrayal of the cause of the Untouchables and the Depressed Classes; and, if I did not believe in this movement before mahatma Gandhi could take it up or Mr. Rajagopalachari went from door to door in Delhi, I should not have been here to move this Bill."

Here was a case of retreat from glory. And what an inglorious retreat? How did Mr. Gandhi react to it? In a statement issued on 4th November 1932, Mr. Gandhi said: "Untouchables in the villages should be made to feel that their shackles have been broken, that they are in no way inferior to their fellow villagers, that they are worshippers of the same God as the other villagers and entitled to the same rights and privileges that the latter enjoy."

"But if these vital conditions of the Pact are not carried out by caste-Hindus, could I possibly live to face God and man? I ventured even to tell Dr. Ambedkar, Rao Bahadur M.C. Raja and other friends belonging to the suppressed group that they should regard me as a hostage for the due fulfilment by caste – Hindus of the conditions of the Pact. The fast, if it is to come, will not be for coercion of those who are opponents of reform, but it will be intended to sting into action those who have been my comrades or who have taken pledges for the removal of Untouchability. If they believe their pledges or if they never meant to abide by them and their Hinduism was a mere camouflage, I should have no interest left in life."

He was never tired of repeating this. Exclusion of the Untouchables from the Hindu Temples, he described, as the agony of his soul. What did Mr. Gandhi do in this connection? Did he resent this betrayal by Mr. Rajagopalachari of this project without which he said he had no interest left in life? One would naturally expect. Mr. Gandhi to denounce this betrayal by the Congress Party to achieve success at the polls? Quite the contrary. Instead of blaming Mr. Rajagopalachari, he blamed Mr. Ranga Iyer for his violent denunciation of the

Congress Party for withdrawing its support to the Bill. This is what Mr. Gandhi said in the issue of the Harijan dated August 31, 1934:

"The ill-fated Temple Entry Bill deserved a more decent burial, if it deserved it at all, than it received at the hands of the mover of the Bill. It was not a bill promoted by, and on behalf of, the reformers. The mover should, therefore, have consulted reformers and acted under instructions from them. So far as I am aware, there was hardly any occasion for the anger into which he allowed himself to be betrayed or the displeasure which he expressed towards Congressmen. On the face of it, it was, and was designed to be, a measure pertaining to religion, framed in pursuance of the solemn declaration publicly made in Bombay at a meeting of representative Hindus, who met under the chairmanship of Pandit Malaviyaji on 25th September, 1932. The curious may read the declaration printed almost every week on the front page of Harijan. Therefore, every Hindu, caste or Harijan, was interested in the measure. It was not a measure in which Congress Hindus were more interested than the other Hindus. To have, therefore, dragged the Congress name into the discussion was unfortunate. The Bill deserved a gentler handling."

The Temple Entry, what one is to say of, except to describe it a strange game of political acrobatics Mr. Gandhi begins as an opponent of Temple Entry. When the Untouchables put forth a demand for political rights, he changes his position and becomes a supporter of Temple Entry. When the Hindus threaten to defeat the Congress in the election, if it pursues the matter to a conclusion, Mr. Gandhi, in order to preserve political power in the hands of the Congress, gives up Temple Entry. Is this sincerity? Does this show conviction? Was the "agony of soul" which Mr. Gandhi spoke of more than a phrase?

Political Demands of Scheduled Castes

Resolutions passed by the Working Committee of the All-India Scheduled Castes Federation held in Madras on the 23rd September 1944 under the Presidentship of Rao Bahadur N. Shiva Raj, B.A.B.L.M.L.A outlining the safeguards for the Untouchables in new constitution.

Resolution No.1

Subject: *Recognition of the Schedule Castes as a separate element.*

The Working Committee of the All India Scheduled Castes Federation has found a section of the Press in India making the allegation, that the statement made by H.E. the Viceroy in his letter to Mr. Gandhi dated the 15th August 1944 to the effect that the Scheduled Castes are one of the important and separate elements in the national life of India and requiring that the consent of the Scheduled Castes to the Constitution of India was a necessary condition precedent for transfer of power to Indians, is a departure from the position of His Majesty's Government as defined in the Cripps Proposals. The committee cannot help expressing its indignation at this propaganda and takes this occasion to state in most emphatic and categorical terms that the Scheduled Castes are a distinct and separate element in the national life of Indian and that they are a religious minority in a sense far more real than the Sikhs and Muslims can be and within the meaning of the Cripps Proposals. The Working Committee desires to point out that what has been stated by Lord Wavell in his letter to Mr. Gandhi has been the position of His Majesty's Government from the very beginning and was enunciated in clear terms as early as 1917 by the authors of the Montagu-Chelmsford Report simultaneously with the enunciation by them of Responsible Government as the goal of India's political evolution and has been confirmed by subsequent action of His Majesty's government such as the grant of separate representation to the Scheduled Castes at the Round Table Conference, Joint Parliamentary Committee and in the Government of India Act, 1935, as a recognized minority, separate from the Hindus.

The Working Committee has, therefore, no hesitation in saying that it is a false and malicious propaganda to allege that this is a departure from the policy of His Majesty's government and regards it as a manoeuvre on the part of the enemies of the Scheduled Castes to defeat their just claims for constitutional safeguards and calls upon Indian political leaders and particularly the Hindu leaders to accept this fact, in the interests of peace and goodwill between the Hindus and the Scheduled Castes, and for the speedy realization of India's political goal.

Resolution No.2

Subject: Declaration by His Majesty's Government relating to the Scheduled Castes and the Constitution

The Working Committee of the All India Scheduled Castes Federation welcomes the declaration made by His Majesty's Government and recently reiterated by His Excellency the Viceroy that His Majesty's Government regards the consent of the Scheduled Castes, among others, to the Constitution of a free India, as a matter of vital importance and as a necessary condition precedent to the transfer of power to Indian hands. At the same time, the Working Committee wishes to draw the attention of His Majesty's Government to the attitude of the Congress and other political organizations in the country which treats this declaration of His Majesty's government as not being a bona fide declaration and made without any intention to honour it and as a mere matter of tactics adopted to postpone transfer of power, and which is in all probability responsible for the unwillingness of the Majority Community to seek for a settlement with the Scheduled Castes. The Working Committee regards this allegation as baseless and calls upon His Majesty's Government not to give any ground for such suspicion and make it clear that they will stand by the declaration at all times and under all circumstances.

Resolution No.3

Subject: Nature of Constitutional Safeguards.

The working Committee declares that no constitution shall be acceptable to the Scheduled Castes unless:

(a) It has the consent of the Scheduled Castes;

(b) It recognizes the Scheduled Castes as distinct and separate element;

(c) It contains within itself provisions for securing the following purposes:

(1) For earmarking a definite sum in the Budgets of the Provincial and Central Governments for the

Secondary. University and Advanced Education of the Scheduled Castes.

(2) For reservation of Government lands for separate settlements of the Scheduled Castes through a Settlement Commission.

(3) For Representation of the Scheduled Castes according to their needs, numbers and importance:

  (i) in the Legislatures,

  (ii) in the Executive,

  (iii) in Municipalities and Local Boards,

  (iv) in the Public Services,

  (v) On the Public Service Commissions.

(4) For the recognition of the above provisions as fundamental rights beyond the powers of the Legislature or the Executive to amend or alter or abrogate.

(5) For the appointment of an Officer similar in status to that of the auditor-General appointed under Section 166 of the government of India Act of 1935 and removable from office in like manner and on the like grounds as a judge of the Federal Court to report on the working of the provisions relating to Fundamental Rights.

Resolution No. 4

Subject: Communal Settlement.

The Working Committee of the All –India Scheduled Caste Federation, while it is most eager for a settlement of the Communal problem, wholly disapproves of the secret negotiations which are being carried on by Mr. Gandhi and Mr. Jinnah for a settlement between the Hindus and the Muslims. The working Committee is of opinion that Communal Settlement of a sectional character is harmful in every way. It is harmful because it ignores the vital interest of other communities. It is harmful because it creates a feeling of suspicion in other communities that dishonest deal is being made between two communities to defeat their interests.

It is also harmful to the general interests of the country, inasmuch as the singling out of one special community from others for conferring special privileges, not necessary for its protection but demanded on the basis of prestige, creates differences in status which from the point of view of maintaining equal citizenship for all, are unjustifiable and must be deplored. The working Committee is surprised that Mr. Gandhi who has time and again proclaimed himself as an opponent of secrecy in public life should have entered into secret diplomacy to bring about Hindu-Muslim settlement. The Committee the communal question, which would give a sense of security and ensure fair and equal treatment to all is to discuss the demands put forth by each interest in public and in the presence of and with the representatives of other interests.

Resolution No. 5

Subject: Revision of the Constitution.

The Working Committee of the All-India Scheduled Castes Federation is of opinion that the provisions in the existing Constitution relating to minority representation are not based on any intelligible principle. The Committee finds that as the system now stands, some minorities have not received even their population ratio of representation, while other minorities have been given weightage over and above their population ratio as a concession to their claims based on historical and military importance. The Working Committee regards the recognition of such claims to be harmful to the interests of other minorities and inconsistent with the ideal of social and political democracy, which is the goal of all Indians and that they should never be tolerated. In this connection, the committee wishes to draw attention to the fact that the principle of giving weightage to specially selected minorities has been condemned by the authors of the Montagu-Chelmsford Report and also by the Simon Commission. The Committee demands that in view of the fact that the next Constitution of India will be for India as a Dominion, the provisions of the Constitution relating to minorities should be revised and should be brought in accord with the principle of equal treatment of all minorities.

Resolution No. 6

Subject: Representation in the Legislatures and in the Executive.

The Working Committee of the all-India Scheduled Castes Federation desires to state in categorical and emphatic terms that the Schedule Castes will not tolerate any discrimination between one community and another in the matter of representation and will insist upon their claim for seats in the Provincial and Central legislatures and in the Provincial and Central Executive being adjudged in the same manner and by the same principles that may be made applicable to the claims of the Muslim community.

Resolution No.7

Subject: Electorates.

The Working Committee of the all-India Scheduled Castes Federation is of opinion that the experience of the last elections held under the government of India Act has proved that the system of joint electorates has deprived the Scheduled Castes of the right to send true and effective representatives to the Legislatures and has given the Hindu Majority the virtual right to nominate members of the Scheduled Castes who are prepared to be the tools of the Hindu Majority.

The working Committee of the federation therefore demands that the system of joint electorates and reserved seats should be abolished and the system of separate electorates be introduced in place thereof.

Resolution No.8

Subject: Framework of Executive Government.

The working Committee of the All-India Scheduled Castes Federation takes note of the fact that not only all wealth, property, trade and industry are in the hands of the Majority community, but even the whole administration of the State is controlled by the Majority Community whose members have monopolized all posts in the State Services both superior and inferior. The working committee of the All-India Scheduled

Castes Federation regards this as the most dangerous situation which cannot but cause great apprehension to the minority communities since the combination of these circumstances gives the majority the fullest power to establish its strangle hold upon the minorities.

The fear of a stranglehold is greatly augmented by the Constitutional provisions relating to the Executive contained, in the government of India Act of 1935 which permits the majority in the Legislature to form a Government without reference to the wishes of the minorities.

The Working Committee of the All –India Scheduled Castes Federation feels that while, in the absence of an alternative system, the Parliamentary system of Government may have to be accepted, the Committee is definitely opposed to the system of Parliamentary cabinet inasmuch as it automatically vests the Executive authority in the Majority Community and thereby strengthens the hold of the Majority Community which has entered into the steel frame of the administration and thus has become a source of great danger to the Minorities.

The Working Committee has, therefore, come to the conclusion that the system of Parliamentary Cabinet is not suited to Indian conditions and that a different system under which Executive Government would be formed in consultation with the wishes of the Minorities must be designed to give them a better sense of security.

The working Committee insists that the Executive in the Provinces well as in the Centre should be constituted in the following manner:

(1) The Executive should consist of a Prime Minister and other ministers drawn from general community and from minority communities in a proportion to be specified in the constitution.

(2) The Prime Minister and Ministers drawn from the general community shall be elected to the Executive by the whole house by single transferable vote.

(3) The Ministers representing the minority communities shall be chosen by the members representing the different communities by single transferable vote.

(4) The Members of the Executive shall be members of the Legislature, shall answer questions, vote and take part in debates.

(5) Any vacancy in the Executive shall be filled in accordance with rules governing the original appointments.

(6) The period for which the Executive shall hold office shall be co-terminus with the life of the Legislature.

Resolution No.9

Subject: Public Services.

While it is desirable to plan for a Government which will be a government of Laws and not of men, it cannot be forgotten that no matter how government is organized, it must remain a Government of men. That being so, whether government is good or bad as distinguished from a merely efficient Government and how far the administration of public affairs will be non political and impartial must depend upon the spirit and outlook and sense of justice of the men who are appointed to administer the Law. The Working Committee of the All-India Scheduled Castes Federation is convinced that the Scheduled Castes can never get protection, justice or sympathy from the present administration which is controlled by men full of caste consciousness, narrow mindedness, absence of sense of justice and having a hatred and contempt for the scheduled Castes. The Working Committee, therefore, demands that the Constitution must recognize the right of the Scheduled Castes to reservation in the Public Services in the same proportion as may be applied to the claims of the Muslim Community.

Resolution No.10

Subject: Provision for Education.

The Working Committee of the All-India Scheduled Castes Federation feels that unless persons belonging to the Scheduled Castes are able to occupy posts which carry executive authority, the Scheduled Castes must continue to suffer, as they have been doing in the past all the injustices and indignities from the hands of the government and the Public. The Working

Kuber, W. N.: *Ambedkar: A Critical Study*, New Delhi, People's Pub. House, 1991.

Lal, Shyam: *Caste and Political Mobilisation: The Bhangis*, Panchsheel Prakashan, Jaipur, 1981.

Lalit K. Sahay: *Ambedkar and Caste System*, Mohit Publications, Delhi, 2010.

Mann, K.: *Tribal Women in a Changing Society*, Mittal Publications, Delhi, 1987.

Marriott, Mckim: *Village India: Studies in the Little Community*, The University of Chicago Press, Chicago, 1955.

Maw, Martin: *Visions of India*, Vertag Perer Lang, London, 1990.

Nath, Trilok: *Politics of the Depressed Classes*, Deputy Publications, Delhi, 1987.

Nim, H.L.: *Thoughts on Dr. Ambedkar*, Siddhartha Educational and Cultural Society, Agra, 1969.

O'Malley, L. S. S.: *Indian Caste Customs*, Vikas Publishing House, Delhi, 1974.

Omvedt, Gail: *Dalits and the Democratic Revolution: Dr. Ambedkar and the Dalit movement in colonial India*, New Delhi, Sage Publications, 1994.

Parvathamma, C. *New Horizons and Scheduled Castes*. Ashish Publishing House, New Delhi, 1986.

Prabhu, P. N.: *Report of the Seminar on Casteism and Removal of Untouchability*, Indian Conference of Social Work, Bombay, 1955.

Pradhan, A. C.: *The Emergence of the Depressed Classes*, Bookland International, Bhubaneshwar, 1986.

Punalekar, S. P.: *Dalit Women in India: Issues and Perspectives*, Gyan Publishing House, New Delhi, 1995.

Raju, G.G.: *The Life of B.R. Ambedkar*, Babasaheb Dr. Ambedkar Memorial Society, Hyderabad, 1979.

Ravindra Prasad Singh: *Ambedkar and Grievances of Scheduled Castes*, ABD Publishers, Delhi, 2010.

Rege, Sharmila: *Sati: A Critical Analysis,* University of Poona, Poona, 1989.

Robb, Peter: *Dalit Movements and the Meanings of Labour in India*, Oxford University Press, New Delhi, 1993.

Rodage, S.: *Subhedar Ramji Ambedkar*. Ramaji Prakashan, Wakvali Dapoli, 1989.

Sangharakshita: *Ambedkar and Buddhism*, Glasgow, Windhorse Publications, 1986.

Shetty, V. T. Rajshekar: *Dalit Movement in Karnataka*, Christian Literature Society, Madras, 1978.

Silverberg, James: *Social Mobility in the Caste System in India*, Mouton Publishers, The Hague, 1968.

Stevenson, Margaret S.: *Without the Pale: The Life Story of An Outcaste*, Association Press (Y.M.C.A.), Calcutta, 1930.

Tope, T.K.: *Dr. B.R. Ambedkar, A Symbol of Social Revolt*, Maharashtra Information Centre, New Delhi, 1964.

Tripathy, R. B.: *Dalits: A Sub-Human Society*, Ashish Publishing House, New Delhi, 1994.

Vakil, A. K.: *Gandhi-Ambedkar Dispute*, New Delhi, Ashish Pub. House, 1991.

Varale, B.H.: *Dr. Babasaheb Ambedkarrancha Sangaki*, Shrividya Prakashan, Pune, 1988.

Index

F

G

H

I

J

K

L

M

N

O

P

R

S

T

U

V

□□□

nmol